Lecture Notes in Computer Science 13962

Founding Editors

Gerhard Goos
Juris Hartmanis

Editorial Board Members

Elisa Bertino, *Purdue University, West Lafayette, IN, USA*
Wen Gao, *Peking University, Beijing, China*
Bernhard Steffen ⓘ, *TU Dortmund University, Dortmund, Germany*
Moti Yung ⓘ, *Columbia University, New York, NY, USA*

The series Lecture Notes in Computer Science (LNCS), including its subseries Lecture Notes in Artificial Intelligence (LNAI) and Lecture Notes in Bioinformatics (LNBI), has established itself as a medium for the publication of new developments in computer science and information technology research, teaching, and education.

LNCS enjoys close cooperation with the computer science R & D community, the series counts many renowned academics among its volume editors and paper authors, and collaborates with prestigious societies. Its mission is to serve this international community by providing an invaluable service, mainly focused on the publication of conference and workshop proceedings and postproceedings. LNCS commenced publication in 1973.

Catherine Dubois · Pierluigi San Pietro

Editors

Formal Methods Teaching

5th International Workshop, FMTea 2023
Lübeck, Germany, March 6, 2023
Proceedings

 Springer

Editors
Catherine Dubois 🆔
ENSIIE
Evry-Courcouronnes, France

Pierluigi San Pietro 🆔
Politecnico di Milano
Milan, Italy

ISSN 0302-9743   ISSN 1611-3349 (electronic)
Lecture Notes in Computer Science
ISBN 978-3-031-27533-3   ISBN 978-3-031-27534-0 (eBook)
https://doi.org/10.1007/978-3-031-27534-0

This Springer imprint is published by the registered company Springer Nature Switzerland AG
The registered company address is: Gewerbestrasse 11, 6330 Cham, Switzerland

Preface

This volume contains the proceedings of the Formal Methods Teaching Workshop, FMTEA 2023. The workshop took place in Lübeck, Germany, on March 6, 2023, affiliated with FM 2023, the 25th International Symposium on Formal Methods.

FMTea 2023 was organized by FME's Teaching Committee. Its broad aim is to support a worldwide improvement in learning Formal Methods, mainly by teaching but also via self-learning. While in recent years formal methods are increasingly being used in industry, university curricula are not adapting at the same pace. Some existing formal methods classes interest and challenge students, whereas others fail to ignite student motivation. It is thus important to develop, share, and discuss approaches to effectively teach formal methods to the next generations. This discussion is now more important than ever due to the challenges and opportunities that arose from the pandemic, which forced many educators to adapt and deliver their teaching online. Exchange of ideas is critical to making these new online approaches a success and having a greater reach.

Previous editions of the workshop were FMTea 2021, held in Porto in October 2019 (https://fmtea.github.io/FMTea19/) and FMTea 2021, held online in Beijing in November 2021 (https://fmtea.github.io/FMTea21/).

The programme committee selected six papers for presentation at the workshop, out of 10 submissions. Each paper was reviewed by at least three referees, and the selection was based on originality, quality, and relevance to the topics of the call for papers. The scientific program contained presentations on various models of teaching, together with innovative approaches relevant for educators of Formal Methods in the 21st century.

The scientific program also included one invited paper and talk given by Erika Ábráham, RWTH Aachen, entitled *Automated Exercise Generation for Satisfiability Checking*. The workshop ended with a special discussion session on the place of Formal Methods in the ACM Curriculum.

We wish to express our thanks to the authors who submitted papers for consideration, to the invited speaker, to the program committee members and to the additional reviewer for their excellent work.

We also thank the EasyChair organization for supporting all the tasks related to the selection of contributions, and Springer for hosting the proceedings. We would like to extend thanks to the organizing committee of FME 2023 for support in all organizational issues. Special thanks to the FME Teaching Committee (website https://fme-teaching. github.io/) and in particular to Luigia Petre, who provided invaluable support and ideas in all phases.

March 2023   Catherine Dubois
                                                                           Pierluigi San Pietro

Organization

Program Chairs

Catherine Dubois   ENSIIE, France
Pierluigi San Pietro   Politecnico di Milano, Italy

Programme Committee

Sandrine Blazy   University of Rennes 1, France
Brijesh Dongol   University of Surrey, UK
Catherine Dubois   ENSIIE, France
Joäo F. Ferreira   INESC-ID & IST, University of Lisbon, Portugal
Alexandra Mendes   University of Porto, Portugal
Claudio Menghi   McMaster University, Canada
José N. Oliveira   University of Minho, Portugal
Luigia Petre   Åbo Akademi University, Finland
Kristin Rozier   Iowa State University, USA
Graeme Smith   The University of Queensland, Australia
Pierluigi San Pietro   Politecnico di Milano, Italy
Emil Sekerinski   McMaster University, Canada

External Reviewer

Nuno Macedo   FEUP & INESC TEC, Porto, Portugal

Invited Speaker

Erika Ábrahám   RWTH Aachen, Germany

Contents

Automated Exercise Generation for Satisfiability Checking

Erika Ábrahám⬤, Jasper Nalbach$^{(\boxtimes)}$⬤, and Valentin Promies⬤

RWTH Aachen University, Aachen, Germany
{abraham,nalbach,promies}@cs.rwth-aachen.de

Abstract. Due to the pandemic, we had to switch our *satisfiability checking* lecture to an online format. To create space for interaction, we gave the students the opportunity to earn bonus points for the final exam by correctly answering some questions during the lecture. It turned out to be challenging to design these questions in a way that makes them relevant, solvable in limited time, automatically evaluated and parametric such that each student gets an individual but comparable variant of the exercise. In this paper, we report the challenges we faced, propose quality criteria for such exercises and discuss these criteria on concrete examples we employed in our teaching.

1 Introduction

Satisfiability checking is a relatively young research area of computer science, aiming at the development of algorithms and tools for checking the satisfiability of logical formulas in a fully automated manner. In the 90's, the urgent need for effective technologies for circuit verification was a driving force for the branch of *propositional satisfiability (SAT)*. Formulas of propositional logic are quantifier-free Boolean combinations of Boolean variables called propositions. The SAT problem poses the question whether a formula of propositional logic is satisfiable, i.e. whether we can substitute truth values for the propositions in the formula such that the formula evaluates to true. For example, the propositional logic formula $(a \lor \neg b) \land (\neg a \lor b \lor c)$ is satisfiable, $a = true$, $b = false$ and $c = true$ being a satisfying assignment. The SAT problem is decidable but NP-complete [6]. Nevertheless, modern SAT solvers are highly efficient on huge real-world problems [15].

Motivated by this success, extensions from propositional logic to first-order logic over different theories have been proposed in the area of *satisfiability modulo theories (SMT)* solving [5]. Considering for example the quantifier-free fragment of real arithmetic, whose formulas are Boolean combinations of polynomial constraints over the reals, the formula $(x - 2y > 0 \lor x^2 - 2 = 0) \land x^4y + 2x^2 - 4 > 0$ is satisfiable, e.g. by the assignment $x = \sqrt{2}$ and $y = 2$.

SAT and SMT solvers enjoy increasing popularity as a wide range of problems from several fields of computer science – including theoretical computer science, artificial intelligence, hardware design and formal verification – can be encoded logically and

Jasper Nalbach was supported by the DFG RTG 2236 *UnRAVeL*.

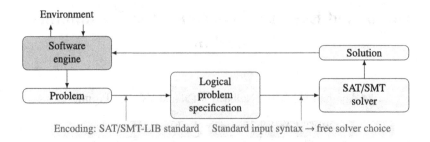

Encoding: SAT/SMT-LIB standard   Standard input syntax → free solver choice

Fig. 1. Embedding SAT and SMT solvers as black-box engines

solved using these technologies, as illustrated in Fig. 1. A major achievement is the establishment of standards [4]: Once a problem is encoded using the standard input language, most tools can be employed directly.

This development requires well-educated experts, who have sufficient knowledge to develop SAT and SMT algorithms and tools as well as to embed them in software technologies effectively. To contribute to this education, since 2008 we offer an elective course on *Satisfiability Checking* for B.Sc. and M.Sc. students at RWTH Aachen University, lasting 13 weeks with 5 h lectures and exercises per week.

The topic is attractive to students, due to its algorithmic nature and its modularity, introducing a number of different algorithms that are related but do not strongly build upon each other, such that missing a lecture or not understanding one topic still allows to continue with the later topics. The number of registered students in the recent years has been 314 (2019), 269 (2020) and 681 (2021). Especially the large 2021 lecture has been very challenging and we decided to restrict the number of student admissions for 2022 to 300, currently having 120 students on the waiting list.

Such large classes in a theoretical course need dedicated pedagogical design, facing the following challenges.

– *Prior knowledge:* Participants who are or were Bachelor students at our university have, when they visit this lecture, already attended lectures on e.g. data structures and algorithms, computability and complexity, mathematical logic, discrete structures, linear algebra and analysis, providing a good basis for the understanding of the theoretical contents. However, Master students who completed their Bachelor studies at other universities often have a different background, which contributes to a diversity in the students' knowledge and skills.
– *Interaction:* Due to the pandemic, but also due to the large number of participants not fitting into our lecture halls, the recent years' lectures have been carried out either online or in a hybrid mode using Zoom. For the 2021 lecture we had about 70 students attending in person, and since the number of online participants has been over 300 we could not use a regular Zoom meeting but had to use a Zoom webinar with extremely restricted opportunities for interaction.
– *Theoretical concepts:* The lecture contents include mathematically quite involved concepts, whose understanding needs active participation. Since direct exchange in form of dialogues is hard to implement in large classes and in Zoom webinars, we needed to find other ways to get the students actively involved.

– *Time schedule:* We offer (annotated) slides, lecture recordings, papers and exercises in our *Moodle* [2] learning platform, but these cannot replace the active participation in the lectures. The lectures are scheduled centrally and in all recent years we had lectures partly at 8:30am, which is an unpopular time for computer science students. Thus we want to provide strong motivation to attend the lectures and offer interaction to help the students to be able to concentrate and to follow complex contents for a typical duration of 90 min.

To address these challenges, an important aspect of good pedagogical design is certainly the inclusion of accessible examples. The theoretical concepts should be motivated and illustrated on problems that all students can understand and relate to, despite their different backgrounds. However, while examples loosen up the lecture and make it easier to follow, they only provide limited possibilities for interaction. They are necessary for a good lecture, but not always sufficient to motivate active participation or even attendance.

Bonus Questions. To overcome the limitations of simply presenting examples, we introduced *bonus questions* which the students can solve during the lecture time to obtain bonus points for the final exam at the end of the semester. For roughly every 45 min of lecture time, after a certain concept or algorithm has been introduced, the lecturer poses a small problem to the students which they can try to solve alone or in groups (in the lecture hall, in Zoom breakout rooms or using other communication channels) for a few (1–3) min. Afterwards, the lecturer presents the solution and the students can ask questions. When all questions have been answered, each registered student gets an individual variant of the presented problem to work on for typically five minutes. For the deployment of the individual tasks to each student and collecting and evaluating the answers, it is practical to conduct the bonus questions online; we used the Moodle learning platform.

The bonus questions promote active participation as they (i) offer a time frame for discussing the presented concepts with other students, and (ii) pose an individual challenge for every student which encourages him or her to question their understanding.

Moreover, the bonus points obtained by solving the questions provide an incentive to attend and actively follow the lectures. In our course, solving the bonus questions offers the students the opportunity to collect up to ten percent of the points in the final exam - a significant amount.

Written Exam. The exam itself entails additional challenges, especially if the pandemic makes in-person exams infeasible. Our last exam was paper-free, conducted *online* using an RWTH-internal tool called *Dynexite* [1]. Similar to Moodle, Dynexite offers the functionality to assign to the students individual questions with different answer types (e.g. multiple choice questions, questions with numerical or textual answers, graphical exercises, etc.), but the evaluation process in Dynexite is specialized to assure compliance with the German exam regulations.

Because the online-scenario severely restricts the options for supervision, it is much harder to prevent students from cheating and sharing answers. It is to some extent possible to use a video conference for controlling that students are present and do not leave their rooms. However, this approach only provides a limited field of vision and a dozen supervisors cannot catch everything happening in an online exam with several hundred participants. In order to counteract the sharing of answers, we created different versions of each task. To ensure fairness, each version should be equally difficult (or easy) to solve. While this means that the tasks are rather similar, it still increases the already high effort needed to correct them. Considering that 380 and 230 students, respectively, registered to our two exams in the year 2021, it is in our interest to keep the correction feasible in a reasonable amount of time.

This is where the online exams provide new opportunities, as they bear excellent conditions for automated processing. They overcome the hurdle of transferring all tasks and solutions to an online system, which is usually too large for in-person exams. Nevertheless, good conditions do not imply that automated correction is trivial. It comes with a number of technical challenges which effectively restrict what kinds of questions can be used in the exams. This means that the requirements imposed by automated processing and online exams should be taken into account already during the creation of the respective tasks.

Practising Exercises and E-Tests. There are further scenarios where automated exercise generation can be employed: First, our students asked for a possibility to re-do the bonus questions for learning purposes. Secondly, as not all students visit the lecture regularly, in order to assure and monitor their progress, we created three obligatory e-tests with 5 exercises each, to be completed within certain deadlines during the semester. At least half of the maximal e-test scores was necessary to be admitted to the exam. As students have several weeks for these sheets and can take them at any time, the temptation to exchange solutions is higher. To mitigate this and to cover more content, the exercises were more complex than the bonus questions.

1.1 Related Work

Publications about automatic exercise generation approaches and tools are rare. Perhaps most closely related to our work is the work in [11] on the automated generation of exercises for online exams in automated deduction (including SAT); the authors faced similar challenges as we did, but they did not conduct an automatic evaluation. In [3,14], the authors report on their experience on generating and evaluating exercises in massively open online courses (MOOCs) for teaching embedded systems; they use SAT/SMT solving for exercise generation and automatic evaluation, however their exercises target more complex practical homework tasks. A tool is presented in [7] for generating exercises for automata and formal languages. Some work has been carried out on generating feedback for programming exercises in [10,17]; however, in these approaches all students get the same exercise.

1.2 An Example: Representing Real Roots

To illustrate the challenges of creating automatically processable bonus or exam questions, we consider an example. Assume we want to check the understanding of the concept of the following representation [12] of *algebraic numbers*:

Definition 1 (Interval Representation). *An* interval representation *(of a real root) is a pair* (p, I) *of a univariate polynomial p with real coefficients and an open interval* $I = (\ell, r) \subseteq \mathbb{R}$ *with* $\ell, r \in \mathbb{Q} \cup \{-\infty, \infty\}$ *such that I contains exactly one real root of p.*

Example 1 (Interval Representation). Using the polynomial $p := x^2 - 2$, we can represent the numbers $-\sqrt{2}$ and $\sqrt{2}$ as $(p, (-\infty, 0))$ and $(p, (0, \infty))$, respectively.

Assume we want to create an exercise to test the understanding of the concept of interval representation. A first, naive approach is to *ask for an interval representation of a real algebraic number*. However, this formulation is too vague for a task that should be quickly solvable online and corrected automatically. Automation requires a uniform input syntax, which needs to be specified and thus makes the question harder to read and to process for the students. Especially in the case of bonus questions, which are usually answered on cell phones, we need to avoid complicated inputs and lengthy questions. Moreover, it is unclear how to evaluate incorrect syntax, since a simple typo should not cause the students to lose all points. Another problem with this formulation is that the answer would not be unique. Not only might this lead to students being uncertain about the answer, it also makes evaluating the task much more complicated. Finally, it might be too easy to guess a correct answer by choosing a large interval.

In the online setting, it would be more appropriate to *ask to select correct representations from a list*, as it does not require a syntax specification and can be answered and corrected conveniently. While this exercise might seem rather simple, it allows to include certain pitfalls which help to make sure that the concept was understood in detail. For example, giving (incorrect) options where the interval is (half-)closed instead of open or where it does not contain exactly one zero of the respective polynomial. Since more answer options also imply a longer processing time for the students, we can reduce the solving effort by using the same polynomial in all options. This way, we obtain tasks like the one below, where we use the polynomial $x^2 - 7x - 8 = (x - 8)(x + 1)$ with real zeros at $x = 8$ and $x = -1$.

Which of the following are interval representations of real roots? (Multiple choice: Please select all interval representations.)

Choose one or more answers:
☐ $(x^2 - 7x - 8, [6, 12])$
☐ $(x^2 - 7x - 8, (-6, -1))$
☑ $(x^2 - 7x - 8, (-7, 5))$
☑ $(x^2 - 7x - 8, (0, 13))$
☐ $(x^2 - 7x - 8, (-10, 16))$
☐ None of the above

---

2 Quality Criteria

Generalizing the observations we made on the example of the interval representation exercise, our objective in this section is to provide general quality criteria for exercise generation. To do so, we recall the side conditions that need to be met by the exercises:

- There is a *limited amount of time* available for solving.
- The answers should be *evaluated automatically*. Due to technical and legal reasons, we need to evaluate each question *all-or-nothing* (i.e. full score is obtained if the answer is fully correct, and no score otherwise).
- The possibilities for *cheat control are limited*. Additionally, we want to encourage collaboration between the students, while testing the individual knowledge.

As seen in the above example, making a "good" exercise can be challenging: The exercise should be suitable to test the understanding of the contents, be individual and fair, but should not cause frustration of students due to lack of time, ambiguous formulation, inappropriate level of difficulty, or technical hurdles when submitting the answer. It is not clear upfront what makes an exercise "good".

We will go through different aspects and propose quality criteria based on our practical experience. We first examine *which problems* are suitable for an exercise, then we see how to *formulate the problem* and how the *answer should be given*, and finally give ideas how to *give feedback* and *present the exercises*.

2.1 Choice of the Problem

The most crucial aspect is the choice of the problem to be solved.

- The problem should be *relevant*:
  - It should address a clearly isolated *central concept* from the lecture, e.g. an important definition, theorem or algorithm.
  - It should *cover basic aspects* of this concept, pointing to different cases or pitfalls.
- The problem should be *solvable*:
  - The *level of difficulty* should be appropriate. This is especially important for bonus questions, as the students should be able to solve them *directly after learning the concept*.
  - It should be *solvable within a given amount of time*.
  - It should be atomic in the sense that *error-propagation is avoided*: if there is more than one non-trivial step, split the test into multiple tests. This way, a miscalculation or misunderstanding of one aspect gives still the chance to prove the understanding of the other aspects. Especially in the setting of the all-or-nothing evaluation, the results are more fair to the students.
  - There might exist different solution paths, but all paths should lead to the same solution.
  - The problem should admit a *short and unique answer* (see also Sect. 2.3).
- The problem should be *parametric*:
  - It should be possible to instantiate the parameters with different values to generate different *problem instances*.
  - The parameter space should admit the generation of a *large amount of instances*. At best, each student gets an individual instance. For bonus tests during lectures, this ensures that each student needs to answer their own question. For homework or (weakly supervised) exams, a large number of instances hardens the collection of catalogues of answers by the students.

- The answers should *not be trivially transferable* between the instances, i.e. it should not be easy to derive the answer for an instance from the answer of another instance.
- The *type of the required computational steps* should be *analogous* for all instances.
- The *number of the required computational steps* should be *comparable* between the instances.
- The *complexity of the required computational steps* should be *comparable* between the instances. For example, it could be meaningful to prefer computations with integers and avoid complex rational numbers or radicals.
- The *size of the problem* should be *scalable*. While bonus tests during the lectures should be minimal, for homework the tasks can be more involved (e.g. cover more cases, involve different types and a larger number of computational steps).

2.2 Formulation of the Problem

Next we turn our attention to the formulation of the problem, i.e. the specification of what the students are expected to do.

- The problem formulation should be as *short* as possible, *easy to understand* and contain only *relevant information*. This especially holds for scenarios where the time available for solving is limited, such that the students can spend most of the time solving the problem, and not on reading and understanding the problem statement.
- The problem formulation should be *clear* and cover all necessary details, so that *no misunderstanding* is possible.
- To make the understanding easier, the *syntax from the lecture* should be followed.

2.3 Answer Format

Once the students solved the problem, we want to know whether their solution is correct. Thus we need to (i) ask a question about the problem's solution and (ii) specify how the answer should be given (syntactically).

- The correctness check of the answers should be *fully automated*. Note that this excludes the listing of long computations as answers.
- The answer should *certify a correct computation*. That means, wrong computations should be detected with high probability.
- Furthermore, the answer space should be large enough such that the *probability to guess the right answer is low*.

Some platforms like Moodle admit several *answer types*, from which an appropriate one should be chosen:

- *Single choice* where multiple possible options or solutions are given, but only exactly one is correct and needs to be selected.

8   E. Ábrahám et al.

- *Multiple choice* is similar to single choice, but multiple options can be selected. Legally, only certain evaluation schemes are allowed in our state (i.e. subtracting points for wrongly selected choices is not allowed). Further, defining a fine-grained evaluation scheme is nearly impossible. For these reasons, we only use multiple choice for questions where all-or-nothing evaluation is reasonable. If all options might be wrong then we introduce *None of the above* as a further option, to avoid gaining scores for doing nothing.
- In *matching* exercises, elements from one set need to be mapped injectively to elements from another set (i.e. fruits to colours).
- Some platforms allow for *graphical* exercises, where one or more points need to be selected on an image. The same remarks hold as for multiple choice.
- *Numerical answers* e.g. for outcomes of computations, or for quantities derivable from a computation. Note that if the number of possible outcomes of a computation is small (e.g. a binary final result) then the answer would be easy to guess; in such cases it is meaningful to ask for some values that describe the computations themselves, instead of just the result.
  If necessary, an input syntax should be defined, e.g. whether non-integers should be entered as reduced fraction or in decimal format with a given number of digits after the comma.
- *Free text* answer fields allow for a great flexibility. The required answer should be *short*, ideally no mistyping is possible. To ease the automated correction, the answers should be *unique*. An input syntax should be specified which allows entering answers also on cell-phones. For lists, a separator should be specified. For math symbols, an encoding for exponents, greek letters, comparison operators, infinity, etc. should be specified. However, even for unique answers, oftentimes several variants of the same answer might be entered (e.g. with or without spaces between list elements). While some input normalization as removal of spaces, conversion to lower case or use of regular expressions might reduce this number, the latter is especially error-prone and it is practically hard to cover all possible answers when stating the exercise. Thus, typically a manual post-correction is necessary.

2.4 Feedback

After the answer is submitted, individual feedback should be given to the students, allowing to learn from their mistakes.

- At least, the *correct answers* should be displayed.
- It is helpful to offer also a description of the *computations* to obtain the answer.
- Optimally, *different verbosity levels* are offered for the explanations.
- For learning, *hints or partial results* could be given on request.

2.5 Presentation of the Exercise

Finally, there are different ways to present the exercises to the students, e.g. with some computer program or app, via a specialized web application, on a learning platform such as Moodle or, in some scenarios, PDF files are sufficient.

- The tool should be *easy to use*, also on *mobile phones or tablets*.
- Formulas and math should be displayed using *LaTeX* for better readability.
- The used platform should be able to *handle sufficiently many simultaneous requests*.

3 Examples

We now illustrate the application of our criteria on four example bonus tests from our satisfiability checking lecture. Each test addresses a central concept and is supposed to be solvable within 5 min. The questions were made available on the Moodle learning platform and the students could answer them using e.g. a computer, a laptop or a cell phone.

3.1 Satisfiability of Logical Formulas

In our first example, we want to test the understanding of the notion of satisfiability for quantifier-free first-order logic formulas. There are several possible problem statements related to this concept, some of which are better suited than others.

- *Ask to provide a satisfiable/unsatisfiable constraint or formula:* This question does not allow parametrization, meaning that students could simply share answers and avoid the individual challenge. Moreover, it allows trivial answers (like $x = x$ for a satisfiable constraint) and therefore does not necessarily indicate how well students understood the concept. Finally, the answer to the question is not unique and requires to fix a format for entering formulas, which makes it harder to understand and to correct automatically.
- *Ask for the set of all solutions of a formula:* This problem can be quite hard to solve by hand, thus taking a long time and being error-prone in the computation steps. In addition, the representation of the solution set might not be unique and requires a fixed (possibly too complex) syntax. It is also not easy to generate comparable, but sufficiently different problem instances, which makes a fair evaluation difficult.
- *Ask to specify some parameters for which a constraint/formula is (un)satisfiable (coefficients, exponents, comparison operators):* While this question is not as hard, its drawbacks are very similar to those of the previous question. The answer and its representation might not be unique and the problem is relatively hard to parametrize and correct fairly.
- *Ask for the satisfiability status of formulas:* As the complexity of the given formulas can be adjusted, the difficulty of this task is scalable to the intended time frame. It can also be easily parametrized by using the same problem structure with different numerical coefficients. The problem can be formulated briefly and clearly, benefiting from the fact that it only allows two possible answers (per formula). The same fact, however, allows to guess the answer too easily if only one formula is given. This can be prevented by asking for the satisfiability status of multiple formulas, for example in a multiple choice question.

We find that the latter problem statement fits the criteria the best and we used it to create questions of the following form:

> Which of the following formulas are satisfiable over the integer domain?
>
> Choose one or more answers:
> ☑ $4x - 1y > 2 \wedge x < 5 \wedge y > 5$
> ☐ $-5u^2 - 3v > 8 \wedge u < 2 \wedge v > 10$
> ☐ None of the above

Each student is given two rather simple formulas constraining two integer variables each, and is asked which of them are satisfiable. A noteworthy detail is the inclusion of the answer "None of the above" to distinguish it from the case that the question was simply not answered. To generate many different variations, for example of the first formula, we used a python script as shown in Listing 1.1.

The first formula has the form $ax - by > c \wedge x < 5 \wedge y > 5$ with two integer variables x and y, and three parameters a, b and c each of which can take on integer values between 1 and 9. This produces 729 different instances with the same simple structure and very much the same level of complexity. We check the satisfiability of each instance using the SMT solver $z3$ [9], and we collect the satisfiable and the unsatisfiable instances in two respective lists. Note that, depending on the parameter domains, the lengths of the two lists might strongly deviate. To avoid right guessing with high probability, for each individual exercise we first randomly select a satisfiability status, and then we randomly select an instance from the corresponding list. Therefore, it is important that we ensure that we produced both satisfiable and unsatisfiable instances.

```
from z3 import *

x = Int('x')
y = Int('y')
s = Solver()

linsat = []
linunsat = []

for c1 in range(1, 10):
    for c2 in range(1, 10):
        for c3 in range(1, 10):
            s.push()
            s.add(c1*x - c2*y > c3, x < 5, y > 5)
            if s.check() == sat:
                linsat.append("\(%i x - %i y > %i \wedge x<5 \wedge y>5 \)"
                                                    % (c1, c2, c3))
            else:
                linunsat.append("\(%i x - %i y > %i \wedge x<5 \wedge y>5 \)"
                                                    % (c1, c2, c3))
            s.pop()

if len(linsat)==0:
    print("No satisfiable linear cases! Exiting, no tests generated")
    quit()
if len(linunsat)==0:
    print("No unsatisfiable linear cases! Exiting, no tests generated")
    quit()
```

Listing 1.1. Python script for exercise generation

We employ a similar method to generate quadratic constraints. Here, it is even more challenging to assure balanced lists of satisfiable and unsatisfiable problem instances. Using integer variables u and v, we instantiate $au^2 - bv > c \wedge u < 2 \wedge v > 10$ with integer parameter values $a \in [-10, -1]$, $b \in [a + 1, 9] \setminus \{0\}$ and $c \in [1, 9]$. Note that we avoid the value 0 for the parameters, as those instances would be easier to solve.

3.2 Tseitin Encoding

A propositional logic formula is in *conjunctive normal form (CNF)* if it is a conjunction of *clauses*, where each clause is a disjunction of potentially negated propositions called *literals*. In the next example, students should prove their capability of applying the *Tseitin encoding* [16], which generates for any propositional logic formula a satisfiability-equivalent formula in CNF. The idea is to recursively replace non-atomic sub-formulas by auxiliary propositions that encode their meaning. For example, for the formula $a \wedge (b \vee c)$ we introduce h_1 to encode the conjunction and h_2 for the disjunction, yielding $(h_1 \leftrightarrow (a \wedge h_2)) \wedge (h_2 \leftrightarrow (b \vee c)) \wedge h_1$. Subsequently, each of the auxiliary encodings can be transformed into three or four clauses, leading to a formula in CNF.

It would be inconvenient for the students to enter entire formulas as answers. Furthermore, the answers would be error-prone and not unique (e.g. in the order of clauses). Instead, we use *matching* exercises, which demand that given, predefined answers are matched to different variations of essentially the same task. The matching method has the advantage that multiple variations can be tested at once, covering different cases and pitfalls of the respective task. Here, we can have one case for each Boolean connective. Moreover, there is no ambiguity in the input format as the answers are selected from a list of predefined options. The students can demonstrate their general understanding without the risk of small errors invalidating their entire answer.

Concerning the automated parametrization of this question, we chose two Boolean operators $\sim_1, \sim_2 \in \{\neg, \vee, \wedge, \rightarrow, \leftrightarrow\}$ and one of two preselected formula structures $a \sim_1 (b \sim_2 c)$ or $(a \sim_1 b) \sim_2 c$, producing 50 different formulas. Choosing five of these formulas and ordering them randomly then generates enough variation for our purpose.

Please assign to the following propositional logic formulas their Tseitin encodings.

$a \wedge (b \vee c)$	$(h_1 \leftrightarrow (a \wedge h_2)) \wedge (h_2 \leftrightarrow (b \vee c)) \wedge h_1$ ▼
$(a \vee b) \rightarrow c$	$(h_1 \leftrightarrow (h_2 \rightarrow c)) \wedge (h_2 \leftrightarrow (a \vee b)) \wedge h_1$ ▼
$a \vee (b \leftrightarrow c)$	$(h_1 \leftrightarrow (a \vee h_2)) \wedge (h_2 \leftrightarrow (b \leftrightarrow c)) \wedge h_1$ ▼
$a \vee (b \rightarrow c)$	$(h_1 \leftrightarrow (a \vee h_2)) \wedge (h_2 \leftrightarrow (b \rightarrow c)) \wedge h_1$ ▼
$a \leftrightarrow (b \vee c)$	$(h_1 \leftrightarrow (a \leftrightarrow h_2)) \wedge (h_2 \leftrightarrow (b \vee c)) \wedge h_1$ ▼

3.3 SAT Solving with DPLL

One of the most important algorithms presented in the lecture is the *DPLL* [8] method for checking the satisfiability of propositional logic formulas in CNF. This algorithm tries to construct a satisfying assignment for the input formula by iteratively choosing a proposition and *deciding* which value to consider for it first, and then applying *Boolean propagation* to detect implications of this decision: if under the current decisions and their implications all literals but one in a clause are *false*, then the remaining last literal must be assigned *true* in order to satisfy the clause and thus the CNF formula. If none of these implications make any clause violated (i.e. all literals *false*) then we choose the next proposition and decide on its value, which might enable new propagations. This continues until either a *full* satisfying assignment is found, meaning that all propositions have a value and all clauses are satisfied, or until the decisions and propagations lead to a situation in which the current assignment assigns *false* to all literals of a clause. This situation is called a *conflict*, which is analysed and resolved by backtracking and reversing decisions.

Example 2 (DPLL). Consider the following propositional logic formula in CNF:

$$(\neg A \vee \neg C) \wedge (\neg C \vee \neg D) \wedge (A \vee B) \wedge (\neg B \vee \neg D)$$

There is no propagation possible in the beginning, so we choose a proposition, e.g. A, and assign *false* to it. Consequently, the third clause propagates $B = true$, leading to $D = false$ by propagation in the fourth clause. Now, all clauses are already satisfied, but C is still unassigned. We simply choose for it the value *false*, giving us a full satisfying assignment.

The complexity and needed time for this task are scalable by choosing more or less complex input sets of clauses. These clause sets can be easily generated automatically and therefore make the problem parametric. For our 5-minutes bonus test, to ensure that all used instances are comparable and simple enough to solve quickly and without many steps prone to error propagation, we use four propositions and clauses with 2–4 alphabetically ordered literals, which gives us 72 different clauses from which we select four randomly. We skip selections where not all 4 propositions are used, or where a clause was selected twice. Furthermore, we apply the DPLL algorithm to each such clause set and consider only those cases for which the algorithm terminates without facing a conflict, and where at least two assignments are fixed by propagation. (However, in the exercise formulation we deliberately do not communicate these facts, since we do not want to provide more information than necessary for a precise specification of the task).

A difficulty of this task is the answer format. Typing in proposition assignments is inconvenient, needs a fixed syntax and must allow different representations. Instead of asking for the actual output of the algorithm, we ask for the *number* of propositions assigned the value *true* at the satisfying assignment or at the first conflict (since we do not announce that no conflict will happen). This leads to a unique and short answer, which still indicates quite reliably whether or not the computations were performed correctly. (Note that we excluded instances where DPLL leads to a conflict not only for comparability, but also because the number of true propositions at the time of a conflict is not necessarily unique but might depend on the order in which we propagate in the clauses).

This results is the following formulation.

Assume the following propositional logic formula in CNF:

$$(\neg A \vee \neg C) \wedge (\neg C \vee \neg D) \wedge (A \vee B) \wedge (\neg B \vee \neg D)$$

Apply the DPLL algorithm until it detects either a conflict or a full assignment. For decisions, always take the smallest unassigned variable in the order $A < B < C < D$ and assign false to it.
At the first conflict or full assignment, how many variables are assigned the value true? Please answer by writing the number using digits without whitespaces.

Answer: ☐ 1 ☐

On the one hand, the question is formulated very precisely, stating the order in which the propositions are assigned, the default value to assign at a decision and specifying that a *full* satisfying assignment is wanted. Also, the required input format is given. This is needed to make the question clear and unambiguous. While a short question is desirable, it should not leave out necessary specifications. On the other hand, we do not need to repeat definitions which were introduced in previous lectures and which the students should know by that point, such as the meaning of a full assignment.

So far, we did not discuss the criteria concerned with *feedback*. The online learning platform Moodle provides the possibility to indicate which answers are correct so that the students receive feedback immediately after they confirmed their answer. There is even the possibility to display a path to the solution, which helps students to understand where they might have made a mistake in the case of a wrong answer. This is particularly important in the case of testing algorithms like DPLL.

Often, the detailed solutions can be generated automatically together with the tasks, though sometimes this requires modifying the respective algorithm to keep track of the data structures and algorithm flow. In the current example, a detailed solution could show the full satisfying assignment and indicate the order in which the assignments were made, as well as information about the propagation used along the way. It is important to keep in mind that a unique solution does not guarantee a unique path to that solution. For instance, we did not specify which proposition should be assigned first in the case that propagation implies the values of multiple propositions. As none of the task's variants lead to a conflict, the order of propagations does not matter for the result, but the paths might differ. Therefore, one should either list all possible solutions, which is more helpful for the students, but also more complicated to implement, or one should indicate that only one of multiple possible solution paths is shown.

3.4 Boolean Resolution

We employed a very similar strategy for creating tests concerning *Boolean resolution*. This concept can be used to resolve conflicts in SAT solving, but we also present it as a stand-alone method for proposition elimination [13].

Assume two clauses $(A \vee l_1 \vee \ldots \vee l_i)$ and $(\neg A \vee r_1 \vee \ldots \vee r_j)$, where A is a proposition and $l_1, \ldots, l_i, r_1, \ldots, r_j$ are literals whose propositions are different from A. From these two clauses we can derive by Boolean resolution the new clause $(l_1 \vee \ldots \vee l_i \vee r_1 \vee \ldots \vee r_j)$, which does not contain A. For a given CNF formula, if we apply this resolution step to all possible pairs of clauses containing A resp. $\neg A$, then the conjunction of all derived clauses, together with the original clauses containing neither A nor $\neg A$, is equisatisfiable to the original formula. Iteratively applying this process, we can eliminate all propositions in a CNF formula and thereby decide its satisfiability: the formula is unsatisfiable if and only if the *trivially false* empty clause has been either part of the input CNF or derived during elimination. Note that during elimination, *trivially true* clauses containing both a proposition and its negation can be neglected.

We treat the above concept of Boolean resolution for satisfiability checking in a dedicated bonus test. As in the last example, we parametrize the task by fixing four propositions; even though it is irrelevant for the satisfiability check, to assure a unique answer, we fix the proposition elimination order. For comparability, for each instance we choose four different clauses with 2–4 alphabetically ordered literals having pairwise different propositions, and admit only instances that refer to at least 3 propositions and for which between 6 and 8 non-trivial clauses are derived. To check whether the students correctly executed the check, we ask for the number of different derived nontrivial clauses that were not part of the input. Note that we need to carefully specify the meaning of *different* to be of semantic nature, independent of the literal order in the clauses.

Apply resolution to eliminate the propositions in the order A, B, C and D from the following propositional logic formula in CNF:

$$(A \vee \neg C \vee D) \wedge (B \vee C \vee D) \wedge (\neg A \vee \neg B \vee D) \wedge (A \vee \neg B \vee D)$$

How many non-trivial (i.e. neither trivially true nor trivially false) clauses are generated during this process that differ in their literal sets from each other and from all input clauses?
Please answer by writing the number using digits without whitespaces.

Answer: | 3 |

---

4   Conclusion

As the pandemic forced us to switch to online teaching, we needed to motivate the students to attend the online lectures and to facilitate interaction between the students in breakout rooms. We looked for ways to add value to live online lectures and the idea of bonus questions came up. As time for development was limited, we had some difficulties at the beginning. It turned out that generating small and relevant parameterized exercises for satisfiability checking is challenging. We iteratively learned and improved our approach, resulting in the list of quality criteria/guidelines we propose in this paper.

We admit that the implementation is tedious and error-prone at first, put it pays off: for the individual tasks as bonus questions, we received very good feedback from students.

Although the pandemic was a trigger for the revision of our teaching concept, we think that these concepts will remain valuable beyond it, thus we plan to further improve our concepts. Not least because of the shift in teaching to flipped classroom concepts where material is provided online which is first processed individually before it is discussed in class, we need material for autonomous learning.

References

1. Dynexite documentation (in German). https://docs.dynexite.de/
2. Moodle. https://moodle.org/
3. Formal methods for engineering education. Technical report UCB/EECS-2015-170, EECS Department, University of California, Berkeley (2015). https://www2.eecs.berkeley.edu/ Pubs/TechRpts/2015/EECS-2015-170.html
4. Barrett, C., Fontaine, P., Tinelli, C.: The satisfiability modulo theories library (SMT-LIB). http://smtlib.cs.uiowa.edu
5. Barrett, C.W., Sebastiani, R., Seshia, S.A., Tinelli, C.: Satisfiability modulo theories. In: Handbook of Satisfiability, Frontiers in Artificial Intelligence and Applications, vol. 185, pp. 825–885. IOS Press (2009). https://doi.org/10.3233/978-1-58603-929-5-825
6. Cook, S.A.: The complexity of theorem-proving procedures. In: Proceedings of the 3rd Annual ACM Symposium on Theory of Computing (STOC 1971), pp. 151–158. ACM (1971). https://doi.org/10.1145/800157.805047
7. D'Antoni, L., Helfrich, M., Kretinsky, J., Ramneantu, E., Weininger, M.: Automata tutor v3. In: Lahiri, S.K., Wang, C. (eds.) CAV 2020. LNCS, vol. 12225, pp. 3–14. Springer, Cham (2020). https://doi.org/10.1007/978-3-030-53291-8_1
8. Davis, M., Logemann, G., Loveland, D.: A machine program for theorem-proving. Commun. ACM 5(7), 394–397 (1962). https://doi.org/10.1145/368273.368557
9. de Moura, L., Bjørner, N.: Z3: an efficient SMT solver. In: Ramakrishnan, C.R., Rehof, J. (eds.) TACAS 2008. LNCS, vol. 4963, pp. 337–340. Springer, Heidelberg (2008). https:// doi.org/10.1007/978-3-540-78800-3_24
10. Gulwani, S., Radiček, I., Zuleger, F.: Automated clustering and program repair for introductory programming assignments. ACM SIGPLAN Not. 53(4), 465–480 (2018). https://doi. org/10.1145/3296979.3192387
11. Hozzová, P., Kovács, L., Rath, J.: Automated generation of exam sheets for automated deduction. In: Kamareddine, F., Sacerdoti Coen, C. (eds.) CICM 2021. LNCS (LNAI), vol. 12833, pp. 185–196. Springer, Cham (2021). https://doi.org/10.1007/978-3-030-81097-9_15
12. Johnson, J.R.: Real algebraic number computation using interval arithmetic. In: Papers from the international symposium on Symbolic and algebraic computation, pp. 195–205 (1992). https://doi.org/10.1145/143242.143311
13. Robinson, J.A.: A machine-oriented logic based on the resolution principle. J. ACM (JACM) 12(1), 23–41 (1965). https://doi.org/10.1145/321250.321253
14. Sadigh, D., Seshia, S.A., Gupta, M.: Automating exercise generation: a step towards meeting the MOOC challenge for embedded systems. In: Proceedings of the Workshop on Embedded and Cyber-physical Systems Education, pp. 1–8 (2012). https://doi.org/10.1145/2530544. 2530546
15. Silva, J.P.M., Lynce, I., Malik, S.: Conflict-driven clause learning SAT solvers. In: Handbook of Satisfiability, Frontiers in Artificial Intelligence and Applications, vol. 185, pp. 131–153. IOS Press (2009). https://doi.org/10.3233/978-1-58603-929-5-131

16. Tseitin, G.S.: On the complexity of derivation in propositional calculus. In: Siekmann, J.H., Wrightson, G. (eds.) Automation of Reasoning. Symbolic Computation, pp. 466–483. Springer, Heidelberg (1983). https://doi.org/10.1007/978-3-642-81955-1_28
17. Wang, K., Singh, R., Su, Z.: Search, align, and repair: data-driven feedback generation for introductory programming exercises. In: Proceedings of the 39th ACM SIGPLAN Conference on Programming Language Design and Implementation, pp. 481–495. ACM (2018). https://doi.org/10.1145/3192366.3192384

Graphical Loop Invariant Based Programming

Géraldine Brieven⍟, Simon Liénardy⍟, Lev Malcev⍟,
and Benoit Donnet⁽✉⁾⍟

Montefiore Institute, Université de Liège, Liège, Belgium
{gbrieven,simon.lienardy,benoit.donnet}@uliege.be,
l.malcev@student.uliege.be

Abstract. This paper focuses on a programming methodology relying on an informal and graphical version of the Loop Invariant for building the code. This methodology is applied in the context of a CS1 course in which students are exposed to several C programming language concepts and algorithmic aspects. The key point in the course is thus to imagine a problem resolution strategy (the Graphical Loop Invariant) prior to writing the code (that becomes, then, reasonably easy once relying on the Graphical Loop Invariant). This paper exposes the rules for building a sound and accurate Graphical Loop Invariant as well as the programming methodology. As such, our programming methodology might be seen as a first step towards considering formal methods in programming courses without making any assumption on students mathematical background as it does not require to manipulate any mathematical notations. The paper also introduces an integrated learning tool we developed for supporting the Graphical Loop Invariant teaching and practice. Finally, the paper gives preliminary insight into how students seize the methodology and use the learning tools for supporting their learning phase.

Keywords: Loop Invariant · Graphical Loop Invariant · Graphical Loop Invariant Based Programming · GLIDE · CAFÉ

1 Introduction

This paper proposes and discusses a graphical methodology, based on the Loop Invariant [13,17], to help students in efficiently and strictly programming loops. This methodology is applied in the context of an Introduction to Programming (i.e., CS1) course alternating between specific C programming language concepts and algorithmic aspects. In particular, the course aims at introducing to first year students basic principles of programming. The concept of a correct and efficient algorithm is highlighted, in the context of a strict programming methodology. Typically, an algorithm requires to write a sequence of instructions that must be repeated a certain number of times. This is usually known as a *program loop*. The methodology we teach for programming a loop is based on an informal version of the *Loop Invariant* (a program loop property verified at each iteration – i.e., at each evaluation of the Loop Condition) introduced by Floyd and Hoare [13, 17]. Our methodology consists in determining a strategy (based on the Loop Invariant) to solve a problem prior to any code writing and, next, rely on the strategy to build the code, as initially proposed by Dijkstra [11].

As such, the Loop Invariant can be seen as the cornerstone of code writing. However, the issue is that it relies on an abstract reflection that might confuse students who may not have the desired abstract background, specially if the Loop Invariant is expressed as a logical assertion. This is the reason why, according to Astrachan [1], Loop Invariants are usually avoided in introductory courses.

That statement is consolidated by much research [20, 22] showing that teaching CS1 is known to be a difficult task since, often, students taking a CS1 class encounter difficulties in understanding how a program works [27], in designing an efficient and elegant program [10] (conditionals and loops have proven to be particularly problematic [8]), in problem solving and mathematical ability [22], and in checking whether a program works correctly [5]. Morever, in our context, due to the large variety of students entering the CS1 program in Belgium[1], we cannot make any assumptions about a first year student's background.

To ensure students follow a strict programming methodology despite their (potential) gaps, we propose a *Graphical Loop Invariant* (GLI). The GLI informally describes, at least, variables, constant(s), and data structures handled by the program; their properties; the relationships they may share, and that are preserved over all the iterations. The goal behind is to generalize what the program must have performed after each iteration. In addition to natural advantages of drawings [15, 25, 26], the GLI allows the programmer to visually deduce instructions before, during, and after the loop. That approach forges abstraction skills without relying on any mathematical background and lays the foundations for more formal methods where the GLI stands as an intermediate step towards a final formal Loop Invariant (being a logical assertion). Our programming methodology is supported by an integrated tool called CAFÉ.

The remainder of this paper is organized as follows: Sect. 2 presents the GLI and how to construct it; Sect. 3 discusses the programming methodology based on the GLI; Sect. 4 introduces the integrated tool for supporting the GLI teaching and practice; Sect. 5 presents preliminary results on how students seize the GLI; Sect. 6 positions this paper with respect to the state of the art; finally, Sect. 7 concludes this paper by summarizing its main achievements and by discussing potential directions for further researches.

2   Graphical Loop Invariant

2.1   Overview

A *Loop Invariant* [13, 17] is a property of a program loop that is verified (i.e., True) at each iteration (i.e., at each evaluation of the Loop Condition). The

---

[1] In which open access to Higher Education is the general rule, with some exceptions.

Loop Invariant purpose is to express, in a generic and formal way through a logical assertion, what has been calculated up to now by the loop. Historically, the Loop Invariant has been used for proving code correctness (see, e.g., Cormen et al. [9] and Bradley et al. [6] for automatic code verification). As such, the Loop Invariant is used "a posteriori" (i.e., after code writing).

On the contrary, Dijkstra [11] proposed to first determine the Loop Invariant and then use it to deduce the code instructions. The Loop Invariant is therefore used "a priori". Our methodology differs from Dijkstra's as we propose to represent the Loop Invariant as a picture: the *Graphical Loop Invariant* (GLI). This picture must depict the variables, constants, and data structures that will appear in the code, as well as the constraints on them; the relationships they may share, and that are conserved all over the iterations. To illustrate the GLI, a very simple problem is taken as example throughout the paper:

Input : Two integers a and b such as a < b
Output : the product of all the integers in [a, b]

Figure 1a shows how the problem should be solved through the corresponding GLI. We first represent the integers between the boundaries of the problem (a and b) thanks to a graduated line labelled with the integers symbol ($\mathbb{Z}$). It models the iteration over all the integers from a to b. Then, to reflect the situation after a certain number of iterations, a vertical red bar (called a *Dividing Line*) is drawn to divide the integer line into two areas. The left area, in blue, represents the integers that were already multiplied in a variable p (p is thus the accumulator storing intermediate results over the iterations). The right area, in green, covers the integers that still have to be multiplied. We decide to label the nearest integer at the right of the Dividing Line with the variable i that plays the role of the iterator variable in the range [a, b]. Of course, the variables i and p must be used in the code. In the following, based on seven rules (see Fig. 2) and predefined drawing patterns (see Sect. 2.3), we provide a methodology for easing the building of a correct GLI (see Sect. 2.2). Then, Sect. 3 details how to deduce code instructions from a GLI manipulation, based on this example.

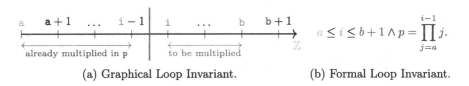

(a) Graphical Loop Invariant.   (b) Formal Loop Invariant.

Fig. 1. Loop Invariant for an integer product between two boundaries.

It is worth noticing here that playing with a graphical version of the Loop Invariant allows students to learn applying formal method programming without manipulating mathematical notations. Indeed, Fig. 1b provides the formal Loop Invariant corresponding to the GLI depicted in Fig. 1a (with the same color code). Producing such a predicate may appear harder to students as it requires to use Mathematical notations (such as $\prod$) with free (i, a, and b) and bound (j) variables. Therefore, the GLI and its usage in code construction might be seen as a first step towards learning and using formal methods in programming.

2.2 Constructing a Graphical Loop Invariant

Finding a Loop Invariant to solve a problem may appear as a difficult task. There are multiple ways to discover a Loop Invariant: e.g., by induction, by working from the precondition, or by starting from the postcondition. For our course, we rather explain to students how to apply graphically the *constant relaxation* technique [14], i.e., replacing an expression (that does not change during program execution – e.g., some n) from the postcondition by a variable i, and use $i = n$ as part or all of the Stop Condition. To help students across that abstract process, we provide seven rules they should apply when searching for a *sound* and *accurate* GLI, as illustrated through Fig. 2. Those rules are categorized into two main categories: (i) *syntax* (i.e., focusing on the drawing aspects only – Rule 1 → Rule 4), (ii) *semantic* (i.e., focusing on the explanations added to the drawings – Rule 1 and Rule 5 → Rule 7).

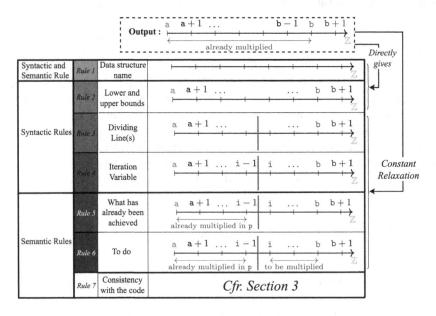

Fig. 2. Designing a GLI step-by-step, from the problem output.

In particular, Fig. 2 shows that it first starts by drawing the program output thanks to a pre-defined pattern (Rule 1 – the different possible patterns are described in Sect. 2.3) and by explaining the program goal (blue arrow and text).

This rule recommends to draw an accurate representation of the data or the data structures relevant for the given problem. Rule 1 also recommends to properly label each drawn data structure (e.g., with variable name). It is essential if several data structures are handled by the program, as they could be mistaken during the code writing. Then, boundaries must frame each structure (Rule 2). Applying Rule 2 prevents some common mistakes, when building the code (see Sect. 3) such as array out of bound errors or overflow. These errors would indeed be more unlikely if the data structure length is properly mentioned in the GLI.

Next, Rules 3 → Rule 6 are sequentially applied in order to roll back the final perspective and visualize the solution under construction through each variable state. Applying Rule 3 makes the Dividing Line appear, naturally reducing the blue zone length (Rule 5) and making room for the green one (Rule 6). The Dividing Lines are the core of our methodology. They symbolize the division between what was already computed by the program and what should still be done to reach the program objective. They enable to graphically manipulate the drawing in order to deduce the code instructions (see Sect. 3).

Applying Rule 4 requires to decide where to place the iteration variable around the Dividing Line, i.e., on the left (thus being part of what has already been achieved by previous iterations) or on the right (as part of the "to do" zone). Usually, we advice students to place it on the right, so that it references the current element to process in the Loop Body. Since a Dividing Line separates what has been done and what is going to be, that means that if we depicted such a line on the data representation as the program is executed, this line would move from one position to another: the first position would correspond to the initial state while the last position would correspond to the final one.

Currently, the application of Rule 5 is partial as it lacks a variable for accumulating intermediate results. It is thus enough to rephrase the sentence below the arrow by introducing the accumulator (i.e., variable p). Applying Rule 5 helps thinking about the behaviour of the program. In order to determine "what has been achieved so far", one should ask the question: "In order to reach the program goal, what should have been computed until now? Which variable properties must be ensured?" Most of the time, this reflection phase highlights either the need for additional variables that contain partial results or relationships between variables that must be conserved throughout the code execution. On the other hand, the information about what has been achieved so far is crucial during the code writing as it helps to decide what are the instructions to be performed during an iteration, i.e., to deduce the *Loop Body* (see Sect. 3).

Finally, it is enough to label the drawing with the "to do" zone right to the Dividing Line, following Rule 6. Naturally, the GLI obtained here is exactly the same as the one provided in Fig. 1a. Applying Rule 6 appears as the less important guideline as it does not bring additional information about the solution. In fact, if we expressed a GLI as a formal one (i.e., as a predicate), there would be no logical notation to describe "what should still be done". Nevertheless, drawing an area indicating what should still be done is a good way to ease the representation of the initial and final states of the program. In the initial state, this "to do area" should span over all the data that is concerned by the program. On the contrary, In the final state, this area should have disappeared while the only remaining area represents what has been achieved by the program. It is then easy to check if the purpose of the program is met in such a state. Moreover, when deducing the code instructions (see Sect. 3), this "to do area" helps to deduce the updates of the variables labelling the Dividing Lines, since the lines have to be moved in order to shrink the area. Finally, it helps finding a Loop Variant to show loop termination as the size of the "to do area" is often a good candidate for the Loop Variant ($b + 1 - a$ on the GLI illustrated in Fig. 1).

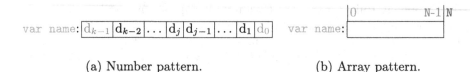

(a) Number pattern.   (b) Array pattern.

Fig. 3. Pre-defined drawing patterns for GLI.

The last rule is a reminder to check if all the variables identified during that reflection phase are actually included in the code.

In addition, to help student identifying the various rules and their applications in a GLI, we adopt a color code (see Fig. 2). This color code is consistent throughout the course and the developed tools (see Sect. 4) and help students understanding exposed GLI during classes.

2.3   Graphical Loop Invariant Patterns

This section introduces some standard patterns for graphically representing common data structures in a CS1 course. Those patterns rely on the first two rules for a correct GLI with the associated color code (see Fig. 2).

Graduated Line. One of the most basic pattern is the graduated line, allowing to represent ordered sets like subsets of Natural or Integers. The line is labelled with the set name (e.g., $\mathbb{N}$ or $\mathbb{Z}$). Each tick on the line corresponds to a value and all those values are offset by the same step. The arrow at the far-right of the line indicates the increasing order of values. That pattern was supporting the GLI presented in the previous subsection and can be seen as resulting from Rule 1 in Fig. 2. Moreover, that line should be framed by boundaries, as performed by applying Rule 2. That directly illustrates the relation $a \leq b$.

Number. For problems concerning a number representation (whether it is binary, decimal, hexadecimal, ...), one can represent this number as a sequence of digits named d_j. The most significant digit is at the right and the least significant one at the left. Often, the d_j are not variables explicitly used in the code but rather figures that, together, represent the actual variable. If a program must investigate the values of the digits in a certain order, it is possible to mention in the picture which is the first and last digits to be handled, as it is, for example, done in Fig. 3a where the least significant digit (in orange) will be used first and the most significant one (in magenta) will be used last.

Array. Figure 3b shows the representation of an array containing N elements. The pattern follows a rectangular shape to depict the contiguous storage of the elements. Above this rectangle, we indicate indices of interest: at least the first (i.e., 0 – always on the left of the drawing, whatever the direction in which the array is processed) and the size N. It is important to see that N is written at the right of the array's border to mean that N is not a valid index as it is out of the array's bounds that are within $[0..N - 1]$. The variable name for accessing the array is written at its left.

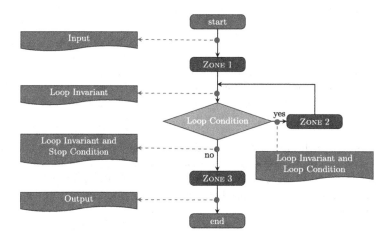

Fig. 4. Loop zones and logical assertions. Blue boxes are block of instructions, orange diamond is an expression evaluated as a Boolean, arrows give an indication of the program flow. Green boxes represent states. (Color figure online)

There are other patterns, such as linked lists and files, but those are usually introduced in a CS2 course in our University.

3   Programming Methodology

Once a GLI meets the rules previously introduced, it can be used to write the corresponding piece of code relying on an iterative process. The general pattern of such a piece of code is given in Fig. 4. Input and Output describe the piece of code input (e.g., a and b such that $a < b$ in the example provided in Sect. 2.1) and the result (e.g., the product of all the integers in $[a, b]$). By definition, a Loop Invariant must be True before evaluating the Loop Condition. The evaluation of the Loop Condition is not supposed to modify the truth value of the Loop

Invariant[2], therefore the Loop Invariant is still True when the Loop Condition is evaluated at True (Loop Invariant and Loop Condition in Fig. 4) or False (Loop Invariant and Stop Condition in Fig. 4). Finally, it is up to the programmer to make sure that the Loop Invariant is True at the end of the iteration, just before the Loop Condition is evaluated, before the potential next iteration.

One can see appearing, in the pattern, four parts that must be filled to form the code: ZONE 1, ZONE 2 and ZONE 3 (standing for instruction(s)) and Loop Condition (standing for a boolean expression). It is worth noting that deducing each part can be done independently, and this, with the help of the GLI.

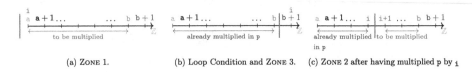

(a) ZONE 1.   (b) Loop Condition and ZONE 3.   (c) ZONE 2 after having multiplied p by i

Fig. 5. Manipulating the GLI for deducing ZONE 1, Loop Condition, ZONE 2, and ZONE 3 for computing the product of integers between a and b. The corresponding GLI is provided in Fig. 1.

To be precise, each part is surrounded, in Fig. 4, by commentaries (in green) that represent conditions that must be satisfied, i.e., be True (e.g., in Fig. 4, ZONE 1 is surrounded by *Input* and *Loop Invariant*). While filling the code of a given part, we must take for granted the information contained in the condition that precedes it and find instructions that will ensure that the condition that follows it is True. The following details these four steps: (*i*) Deducing variables initialisation (ZONE 1) from the drawing of the initial state; (*ii*) Deducing the Stop Condition (and thus the Loop Condition) from drawing the final state; (*iii*) Deducing the Loop Body (ZONE 2) from the GLI; (*iv*) Deducing the instructions coming after the loop (ZONE 3) from drawing the final state. These four steps can be achieved in any order, except the Loop Body determination that may require to know the Loop Condition. Both initial and final states are obtained from the GLI through graphical modifications. Those steps are detailed below, illustrated by the example introduced in Sect. 2.1.

It is worth noting that Fig. 4 and the various zones pave the way for a more formal approach in code construction that relies on Hoare's triplet [13,17], with respect to a strongest postcondition code construction approach [11]. For instance, the Loop Body (i.e., ZONE 2) may be seen as {INV ∧ B} ZONE 2 {INV}, where INV stands for the formal Loop Invariant and B for the Loop Condition. In addition, graphical manipulation of the GLI corresponds to logical assertions describing states between instructions.

ZONE 1. First, the GLI provides information about the required variables. In our example, we need four variables: a, b, i, and p. a and b are provided as input

---

[2] To make it simple, we do not consider here side effect expressions, e.g., pre- or post-increment.

to the piece of code. It is worth noticing that the drawing provides a clue about
the variables type: they are all on a graduated line labelled with $\mathbb{Z}$, meaning
they are of type int.

The initial values of the variables can be obtained from the GLI by shifting
the Dividing Line (in red) to the left in order to make the blue zone (i.e., the zone
describing what has been achieved so far by previous iterations) disappear. The
variable labelling the Dividing Line (i) is also shifted to the left accordingly and
stays at the right of the Dividing Line. By doing so, as seen in Fig. 5a, the initial
value of i must be a (i.e., the particular value just below i in Fig. 5a). With
respect to the variable p, we know from the GLI (see Fig. 1a) that it corresponds
to the product of all integers between a and the left-side of the Dividing Line (i.e.,
i-1). As this zone is empty, we deduce the initial value of p as being the empty
product, i.e., 1. The following piece of code sums up the deduced instructions
for ZONE 1:

```
int i = a;
int p = 1;
```

Stop Condition and Loop Condition. Determining the Loop Condition
requires to draw the final state of the loop, i.e., a state in which the goal of
the loop is reached. Since the purpose of our problem is to compute the product
of the integers between a and b, we can obtain such a representation from the
GLI (Fig. 1a) by shifting the Dividing Line (in red) to the right, until the green
zone (i.e., "to do" zone) has totally disappeared. In the fashion of ZONE 1, the
labelling variable i is shifted at the same time as the Dividing Line. This graphical manipulation leads to Fig. 5b where we can see that the goal of the loop is
reached when $i = b + 1$ and the iterations must thus be stopped. The loop Stop
Condition is therefore $i = b + 1$. As the Loop Condition is the logical negation
of the Stop Condition, it comes $i \neq b + 1$. However, We recommend, in order to
properly illustrate the relationship between i and b, to use a stronger condition,
i.e., $i < b + 1$ or $i \leq b$ that is, of course, equivalent. The following piece of code
sums up the deduced instructions for the Loop Condition:

```
while(i <= b)
```

ZONE 3. As we just depicted the final state (see Fig. 5b), we can see that the
variable p holds the product of the integers between a and b, meeting the program
goal. Nothing remains to be done after the loop in this case. However, ZONE 3
is not necessarily empty. For example, in a program that computes the average
of a certain numbers of values, the loop would sum and count the values and
ZONE 3 would be the division of the sum by the number of counted values.

ZONE 2. Determining the Loop Body is often the most difficult step. We start
from what we know: both the Loop Condition and the GLI are True (See the general loop pattern in Fig. 4). We must find instructions such that it will progress
the situation towards the program goal. In other words, make the blue zone

increase and the green zone decrease. As the blue zone represents the integers that are already multiplied in p (thus from a to i-1), we can make this zone grow by multiplying the next integer to p. This next integer is read in the GLI at the right of the Dividing Line: i.

After having multiplied p by i, the situation in Fig. 5c is obtained. It must be noted that is not the GLI anymore since the variable i is now at the left of the Dividing Line. In this particular situation, the GLI is False, whatever the particular values of a, b, i, or p. According to the loop pattern (See Fig. 4), we must recover the GLI, i.e., make it True again, before the end of the Loop Body. By comparing the Fig. 1a and Fig. 5c, we can see that in the GLI, the value labelling the right side of the Dividing Line is i and in the current situation, this is i +1. Therefore, by assigning the value i +1 to i (i.e., increasing i), the GLI is restored. Finally, the following piece of code shows the Loop Body instructions:

```
p *= i;
i++;
```

4   Learning Tools

This section describes how students can practice the GLI. The main goal is to provide students with a structured and coherent framework so that they do not start their loop design from scratch. To meet that purpose, as early stage, the Blank GLI method is proposed through the *Programming Challenge Activity* (PCA) [18]. The Blank GLI provides a canvas to students. That canvas frames students' solutions so that the semantic of a given solution can be automatically corrected and commented. That GLI correction is handled through a tool called CAFÉ [19][3]. Besides this, CAFÉ also supports GLIDE, a sketching module dedicated to the GLI. Those different components are detailed below and their interaction is illustrated in Fig. 6.

GLIDE. The *Graphical Loop Invariant Drawing Editor* (GLIDE) guides students in drawing their GLI by using the predefined graphical patterns (see Sect. 2.3) and following the first six rules (Sect. 2.2). GLIDE is illustrated in Fig. 7. On the top-left, you can notice a drop-down list itemizing the different drawing patterns. Once a student has selected the appropriate one, they can start formally

---

[3] The version of CAFÉ discusses in this paper corresponds to an upgrade with respect to Liénardy et al. [19].

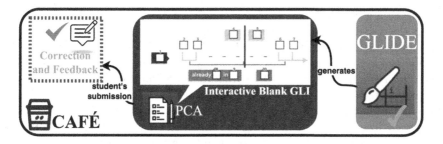

Fig. 6. Link between the Blank GLI, GLIDE, and CAFÉ.

describing their loop mechanism according to the first six rules. The graphical components the student can use are available on the left. Each of them is mapped to one/several rule(s).

Once a student considers their GLI is completed, they can submit it and some basic checks are performed. In particular, syntactic mistakes are detected (such as the lowerbound being further than the upperbound or some description of what has been achieved so far that is missing). However, the GLI semantic is not verified, which means that the solution can be positively assessed by the GLIDE while the GLI does not make sense.

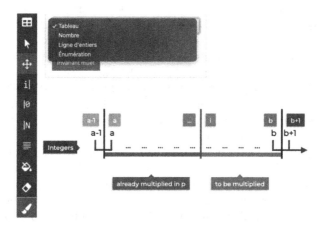

Fig. 7. Screenshot of GLIDE.

Interactive Blank GLI. The Blank GLI consists in providing a canvas the students have to fill out. The Blank GLI corresponding to the example introduced in Sect. 2.1 is illustrated as part of Fig. 6. Such a blank drawing depicts only the general shape a correct and rigorous GLI should follow (i.e., partially Rule 1) in response to a given problem. Students must then annotate properly the canvas so that the drawing becomes the figure of their Loop Invariant for their solution to the particular problem to be solved.

Any Blank GLI always comes with two types of box: (*i*) red boxes standing to host expressions (i.e., constants, variables, operations, or left blank) and are to be completed by students without support; (*ii*) green boxes standing to host labels that students must drag and drop from a pre-defined list. That list contains multiple choices, some of them being the expected answers, others being purely random. Doing so, we pave the way for an automatic correction of the GLI (with strong feedback). This can be achieved thanks to the fact each box is numbered. In this way, when a student's solution gets corrected, each piece of solution is easy to be pointed out, allowing to bring a rich feedback while still keeping it clear and smooth to digest for the student. That process is supported by CAFÉ [19].

Programming Methodology with CAFÉ. CAFÉ [19] is a tool we initially developed in order to support a remote programming activity (PCA) [18]. CAFÉ's purpose is to correct students' work and provide instantaneous personalized feedback and feedforward, based on their mistakes. Their mistakes are mapped to error codes classified in a misconception library. That library has been fed based on previous experiences. Some error codes are defined for each step and each zone of the GLI. They also cover most of the inconsistencies that may occur between the GLI and the resulting code to make sure the student really utilizes the methodology. Also, it is worth noticing CAFÉ gives the opportunity to catch students' learning behavior by collecting data.

That correction and feedback scope got extended after having led an assessment of CAFÉ's impact on students' learning [7]. Now, CAFÉ embeds GLIDE and offers a friendly interface to students when they are solving a given problem. That interface (illustrated in Fig. 8) structures and sequences the construction of the solution, aligned with the programming methodology (Sect. 3). In particular, Fig. 8 adresses the problem consisting in compressing a given array, based on consecutive elements whose sum is 10.

On Fig. 8, one can see that the GLI and the code are represented through successive frames. By taking a closer look, it can be noticed that the GLI (in the upper frame) is divided into four tabs, one for the Blank GLI, one for building and justifying (by moving the Dividing Line) ZONE 1, one for building and justifying (by moving the Dividing Line) the Stop Condition, and for the Loop Variant. It is important to notice that a student cannot access the next tab if the current one has not been filled. That locking path approach also applies at a higher level, between the GLI and the coding steps. That feature aims to impose students to sequentially follow the methodology and not directly jump to the code without any proper design to rely on.

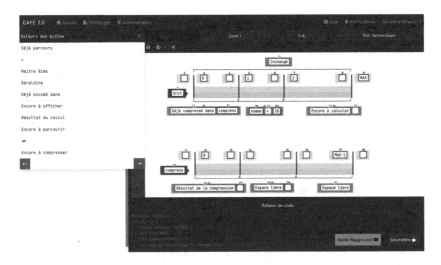

Fig. 8. Interface supporting the programming methodology.

5   Preliminary Evaluation

We surveyed students ($N = 70$), after the final exam, with the question "*What drives/discourages you in using the* GLI?". 36% of students highlighted the method difficulty, limiting so the advantages of the programming methodology. Then, 30% of students were convinced the methodology is useless and directly coding is manageable. This last opinion may suggest that the problems difficulty exposed to students should be increased, so that the importance of the program methodology would be better highlighted. However, a balance must be found between exercises difficulty and methodology mastering, which requires starting with easy problems. An alternative is to enforce the guidance over the GLI construction (like CAFÉ does in its most recent version), so that harder problems can be provided while remaining accessible.

With respect to GLI construction and tools usage, we can show how much the blank GLI (practiced through the PCA) and GLIDE can be relevant in students learning journey. First, there is a correlation between students' exam grades and students' participation to the PCA ($r = 0.57$, $p < 0.0001$). The GLI approach (supported by the PCA) seems thus to forge students' ability to construct a correct and sound GLI from scratch (which is what students are expected to perform in the exam). This inference gets corroborated by students' opinion collected through another survey ($N = 79$) addressed the year before. More precisely, from the statement "*The Blank* GLI *is useful to find out the* GLI.", 47.4% of students agreed or strongly agreed on, 24.4% disagreed and 25.3% standed in between.

In addition, we looked at the possible correlation between exam grades and GLIDE usage ($r = 0.42$, $p < 0.0001$). The lower impact of GLIDE compared to the Blank GLI may be due to the fact that some students still lack landmarks in using GLIDE while the Blank GLI frames more students' solution. Now that the guidance has been enforced in CAFÉ's last version, we expect to see students reaching an upper step and being able to better take advantage from their experience with GLIDE. That premise is subject to future work.

6   Related Work

While there is an abundant literature on Loop Invariants for code correctness and on automatic generation of Loop Invariants (see for instance [9] or Bradley et al. [6]), their use for building the code has attracted little attention from the research community. With respect to Loop Invariant based programming, the seminal work has been proposed by Dijkstra [11], followed by Meyer [23], Gries [16], and Morgan [24]. As such, the program construction becomes a form of problem-solving, and the various control structures are problem-solving techniques. Those works proposed Loop Invariants as logical assertions.

Tam [28] suggests to introduce students to Loop Invariant as early as possible in their courses and describes several examples of code construction based on informal Loop Invariants expressed in natural language. Astrachan [1] suggests the use of Graphical Loop Invariants in the context of CS1/CS2 courses. However, his approach is incomplete as the suggested drawing lack of completeness (e.g., objects manipulated, such as arrays, are not named in the drawing), might lead to confusion (e.g., variables positions around the dividing line are somewhat unclear), and the drawing is not explicitly manipulated to derive particular situations. Back [2–4] proposes nested diagrams (a kind of state charts) representing, at the same time, the Loop Invariant and the code. However, in such a situation, Loop Invariants are expressed as logical assertions. Since, Manilla [21] has evaluated the impact of errors in those nested diagrams. Finally, Erkisson et al. [12] propose a pictorial language for representing Loop Invariants. Their language only applies to arrays and is a mix between drawings (the data structure is drawn and partitions are colored to illustrate universally quantified predicate) and formal languages (the meaning of partitions is expressed as a predicate).

7   Conclusion

This paper introduced a GLI based programming methodology consisting in depicting a graphical representation of the Loop Invariant to solve a given problem prior to writing any piece of code. This methodology is currently taught in a CS1 course. Some preliminary results showed that many students cannot embrace it, mainly because they do not perceive its interest and they miss abstraction skills. Seing that, when we define a problem, a tradeoff must be found between its complexity (so that students feel the purpose of the methodology) and its accessibility (so that students are able to solve it). To reconcile those characteristics, CAFÉ was proposed as an integrated learning tool supporting the methodology and guiding students in solving more complex statements. In particular, a resolution framework is provided as well as personalized feedback so that students are able to refine their understanding. Besides this, it enables more transparency about individual students' learning behavior and resulting performance on the GLI thanks to collected data.

In future work, it is planned to harness that data to accurately assess the methodology by closely analysing students' learning path towards mastering the GLI. In particular, we will capture how much time students spend on the GLI and the code, respectively to see if they put their effort on the GLI. We will also track how students construct their solution to confirm they follow the steps suggested by CAFÉ. Finally, a focus will be dedicated to the way students read, integrate, and take advantage of the feedback to improve their skills in constructing a GLI. Besides that deeper analysis on the GLI, it is aimed at formalizing the translation from the GLI into a logical assertion in order to end the bridge towards a formal method (being the Loop Invariant here).

References

1. Astrachan, O.: Pictures as invariants. In: Proceedings of the Twenty-Second SIGCSE Technical Symposium on Computer Science Education, pp. 112–118 (1991)
2. Back, R.-J.: Invariant based programming. In: Donatelli, S., Thiagarajan, P.S. (eds.) ICATPN 2006. LNCS, vol. 4024, pp. 1–18. Springer, Heidelberg (2006). https://doi.org/10.1007/11767589_1
3. Back, R.J.: Invariant based programming: basic approach and teaching experiences. Form. Asp. Comput. **21**(3), 227–244 (2009). https://doi.org/10.1007/s00165-008-0070-y
4. Back, R.J., Eriksson, J., Mannila, L.: Teaching the construction of correct programs using invariant based programming. In: Proceedings of the 3rd South-East European Workshop on Formal Methods (SEEFM) (2007)
5. Ben-David Kolikant, Y., Mussai, M.: "So my program doesn't run!" Definition, origins, and practical expressions of students' (mis)conceptions of correctness. Comput. Sci. Educ. **18**(2), 135–151 (2008)
6. Bradley, A.R., Manna, Z.: The calculus of computation: decision procedures with applications to verification. Springer, Heidelberg (2007). https://doi.org/10.1007/978-3-540-74113-8
7. Brieven, G., Liénardy, S., Donnet, B.: Lessons learned from 6 years of a remote programming challenge activity with automatic supervision. In: Väljataga, T., Laanpere, M. (eds.) Shaping the Digital Transformation of the Education Ecosystem in

Europe. EDEN 2022. CCIS, vol. 1639, pp. 63–79. Springer, Cham (2022). https://doi.org/10.1007/978-3-031-20518-7_6

8. Cherenkova, Y., Zingaro, D., Petersen, A.: Identifying challenging CS1 concepts in a large problem dataset. In: Proceedings of the 45th ACM Technical Symposium on Computer Science Education, pp. 695–700 (2014)

9. Cormen, T.H., Leiserson, C.E., Rivest, R.L., Stein, C.: Introduction to Algorithms. MIT Press, Cambridge (2009)

10. Dale, N.B.: Most difficult topics in CS1: results of an online survey of educators. ACM SIGCSE Bull. **38**(2), 49–53 (2006)

11. Dijkstra, E.W.: A Discipline of Programming. Prentice-Hall Inc., Englewood Cliffs (1976)

12. Eriksson, J., Parsa, M., Back, R.-J.: A precise pictorial language for array invariants. In: Furia, C.A., Winter, K. (eds.) IFM 2018. LNCS, vol. 11023, pp. 151–160. Springer, Cham (2018). https://doi.org/10.1007/978-3-319-98938-9_9

13. Floyd, R.W.: Assigning meanings to programs. In: Proceedings of the Symposium on Applied Mathematics (1967)

14. Furia, C.A., Meyer, B., Velder, S.: Loop invariants: analysis, classification, and examples. ACM Comput. Surv. **46**, 1–51 (2014)

15. Ginat, D.: On novice loop boundaries and range conceptions. Comput. Sci. Educ. **14**(3), 165–181 (2004)

16. Gries, D.: The science of programming. In: Monographs in Computer Science. Springer, New York (1987). https://doi.org/10.1007/978-1-4612-5983-1

17. Hoare, C.A.R.: An axiomatic basis for computer programming. Commun. ACM **12**(10), 576–580 (1969)

18. Liénardy, S., Leduc, L., Verpoorten, D., Donnet, B.: Challenges, multiple attempts, and trump cards: a practice report of student's exposure to an automated correction system for a programming challenges activity. Int. J. Technol. High. Educ. **18**(2), 45–60 (2021)

19. Liénardy, S., Leduc, L., Verpoorten, D., Donnet, D.: CAFÉ: automatic correction and feedback of programming challenges for a CS1 course. In: Proceedings of the ACM Australasian Computing Education Conference (ACE), pp. 95–104 (2020)

20. Luxton-Reilly, A., et al.: Introductory programming: a systematic literature review. In: Proceedings Companion of the 23rd Annual ACM Conference on Innovation and Technology in Computer Science Education (ITiCSE), pp. 55–106 (2018)

21. Manilla, L.: Invariant based programming in education – an analysis of student difficulties. Inform. Educ. **9**(1), 115–132 (2010)

22. Medeiros, R.P., Ramalho, G.L., Falcão, T.P.: A systematic literature review on teaching and learning introductory programming in higher education. IEEE Trans. Educ. **62**(2), 77–90 (2019)

23. Meyer, B.: A basis for the constructive approach to programming. In: IFIP Congress, pp. 293–298 (1980)

24. Morgan, C.: Programming from Specifications. Prentice-Hall, Hertfordshire (1990)

25. Nilson, L.B.: The Graphic Syllabus and the Outcomes Map: Communicating Your Course. Wiley, San Francisco (2009)

26. Pólya, G.: How to Solve It. Princeton University Press, Princeton (1945)

27. Schröter, I., Krüger, J., Siegmund, J., Leich, T.: Comprehending studies on program comprehension. In: Proceedings of the IEEE/ACM International Conference on Program Comprehension (ICPC), pp. 308–311 (2017)
28. Tam, W.C.: Teaching loop invariants to beginners by examples. In: Proceedings of the ACM Technical Symposium on Computer Science Education (SIGCSE), pp. 92–96 (1992)

A Gentle Introduction to Verification of Parameterized Reactive Systems

Nicolas Féral[1] and Alain Giorgetti[1,2]

[1] Université de Franche-Comté, 25000 Besançon, France
[2] Université de Franche-Comté, CNRS, institut FEMTO-ST, 25000 Besançon, France
`alain.giorgetti@femto-st.fr`

Abstract. An introduction to symbolic model-checking and deductive verification techniques is offered to Master's students at the University of Franche-Comté. This teaching is carried out remotely. It is built around the use of the Cubicle model-checker and the Why3 platform. It shows how to verify the safety of distributed reactive systems which are parameterized by the number of processes run in parallel, when this safety is expressed as non-reachability of critical states.

For remote lab sessions, a virtual machine containing the Cubicle software and the Why3 platform is provided to the students. Examples of reactive systems are specified by the students in the input language of Cubicle and in the WhyML language of Why3. With Cubicle, students use a backward reachability algorithm that discovers dangerous states of these systems by tracing back their transitions to critical states. Safety is proven if no initial state is reached. When a system is safe, Cubicle produces a certificate in WhyML, which contains an invariant synthesized by Cubicle. This certificate can be executed with Why3 to prove that the system indeed preserves this invariant. Students also learn how to directly specify reactive systems in the WhyML language, using a primitive for non-determinism between transitions and between processes that evolve inside each transition.

Keywords: Formal methods · Parameterized reactive systems · Safety · Reachability · Why3 · Cubicle

1 Introduction

This article presents a course taught at the University of Franche-Comté, entitled *Specify and Verify*, which allows students to discover the notion of reactive system and to use an implementation of a technique of symbolic model checking, in order to verify the safety of parameterized reactive systems. The parameterization of these systems relates to the number of identical processes which run in parallel. Safety is expressed here as simply as possible, as the unreachability of critical states. The challenge is to find an invariant of the global system which excludes all the critical states. The verification then consists in proving – preferably in an automatic way – that all the initial states of the system satisfy the invariant, that all its transitions preserve it, and that no critical state satisfies it. The objective being to verify systems of any size, independently of the number of processes, the applied technique is *symbolic model-checking*, here based on representations of states and transitions in first-order logic. In this course, the description of parameterized reactive systems and the verification of their safety use the *model checker* Cubicle [3] and the deductive verification platform Why3 [2].

This course is part of a curriculum that has been designed for many years to be delivered entirely by distance learning. It is primarily intended for students who cannot attend face-to-face classes, for various professional or personal reasons. Students must study at home, alone and at different times. The main pedagogical objective is that the students become able to apply to simple systems a process of *formal specification* and *computer-assisted verification* of the consistency of the specifications.

In addition to the verification of reactive systems, this course also covers functional specification and deductive verification of simple imperative programs, such as a function calculating the factorial or performing a search for elements in an array. This article does not address this subject, which is more classical than the verification of parameterized reactive systems, and already covered in other articles, for example [1].

This article is written by A. Giorgetti, teacher at the University of Franche-Comté, manager and tutor of the course *Specify and Verify* since 2018, and N. Féral, student of this module in 2020–21. The graduation project of N. Féral, about automated verification of parameterized reactive systems, has greatly contributed to the introduction of Cubicle in this course, from September 2021. The educational material presented in this document can be downloaded from the professional web page https://members.femto-st.fr/alain-giorgetti/en of the second author.

Section 2 situates this course in the master curriculum and describes its main characteristics, then its pedagogical progression. Section 3 details the notions covered during the course. Section 4 describes the working environment provided to the students, in order to assimilate the course. Course evaluation procedures are discussed in Sect. 5. Finally, Sect. 6 shares remarks and considerations inspired by the design and development of this course.

2 Description of the Teaching Unit

The *Specify and Verify* module is part of the last semester of the *advanced computing and applications* and *software development and validation* tracks of the computer science master's degree at the University of Franche-Comté. Both tracks are entirely delivered remotely. They benefit from a long experience and recognized know-how of the University of Franche-Comté in distance education, since 1966, for the preparation and delivery of national university diplomas. All teaching infrastructure is digitally managed and accessible online, using the Moodle e-learning platform (https://moodle.org).

The main Moodle page for the course *Specify and Verify* provides access to written course material, exercises and homework sheets, corrected annals, but also to a discussion forum between students and with teachers. The latter provide regular tutoring, answering students' questions on this forum and by e-mail. Students are strongly encouraged to respect a provided study schedule, similar to the content of Table 1.

Table 1. Pedagogical progression of the course

Week	Type	Topic
1	Lesson 1	Introduction to model-based verification and proof of programs
1	Exercises	Lesson 1 assimilation exercises
2	Lesson 2	Specification and verification of parameterized reactive systems with Cubicle
2	Homework 1	Creation of the lab environment (with Docker), specification and verification of a first reactive system
2	Exercises	Exercises related to Lesson 2: specification and verification of examples of parameterized reactive systems with Cubicle
3	Lesson 3	Logic of Why3 (propositional and first-order logic with basic and inductive types), first contact with this platform
3	Exercises	Assimilation of the logic of Why3: propositions, quantifiers, predefined types, formalization of simple problems and properties
4	Deepening	Study of the personalized correction of the homework 1 and its provided solution
4	Lesson 4	Specification and deductive verification of simple imperative programs in WhyML language
4	Exercises	Exercises related to Lesson 3
4	Homework 2	Study of WhyML certificates generated by Cubicle, verification of imperative programs with Why3
5	Revisions	Deepening and preparation for the final exam, with the help of provided annals
6	Deepening	Study of the personalized correction of the homework 2 and its provided solution

The course consists of 4 lessons. The first one is an introductory chapter that places the model verification approach within the context of the software and system development process. It distinguishes between declarative models (specifications) and operational models. It defines and distinguishes reactive systems, open or closed, which mainly interact with their environment, and transformational systems, which carry out a calculation. This introduction also defines the methods of model-checking and deductive verification, more commonly called "program proof". The second lesson presents the model checker Cubicle, its specification language and its methods to verify the safety of a parameterized reactive system. The third lesson presents the deductive verification platform Why3 and its input language, named WhyML. The fourth lesson teaches the proof of (small) imperative programs specified and implemented in the WhyML language.

The exercises allow the students to assimilate the lessons, by looking in the course material for the relevant notions for their resolution, and the tasks to be carried out to achieve the required objectives. Each exercise sheet is associated with a lesson of the course, as detailed in Table 1. A solution of the exercises is distributed separately, sometimes after a small delay, to encourage students to solve the exercises without consulting it. Thus, the study of an exercise sheet corresponds to a tutorial session carried out remotely.

A homework subject is made up of exercises, sometimes a little more exploratory than the course's assimilation exercises. Homework assignments must be returned within a set deadline. Then, each student's homework is marked and annotated by the tutors, and returned with a standard solution and a detailed scale, so that the students can learn from their mistakes, identify their difficulties and measure their progress. In the first homework subject, the first exercise helps the students to set up the working environment, as detailed in Sect. 4. A second exercise proposes to the students to model and verify with Cubicle a reactive system described in natural language. The second homework subject is dedicated to the use of the Why3 platform. In a first exercise, the study of the system modeled during the first homework is completed by the generation with Cubicle of a certificate of proof of safety for the Why3 platform. Students should be able to identify the different parts of the certificate, in particular the invariant synthesized by Cubicle and the logical goals for its preservation proof. In a certificate, each transition is formalized by a logical relation between any state s of the system before the transition and any state s' after the transition. This primed notation for states after the transitions is also used to define by a primed predicate I' the invariant I after the transitions. Thus, reading certificates introduces students to the before-after relational semantics of action systems, and to a definition of the notion of inductive invariant formalized in first-order logic. A second exercise can require the programming of a reactive system in WhyML, then the design and the realization of a proof of an invariant for this system. In addition to reactive system verification, the second homework may also contain imperative program verification exercises.

3   Teaching Content

This section details the essential notions on the modeling and verification of systems studied in the module *Specify and Verify*, and then the educational documentation provided for the use of Cubicle and Why3 when carrying out the practical questions of the exercises and homeworks.

3.1   Taught Concepts

The course distinguishes between transformational systems, which calculate results from data and according to an algorithm, and reactive systems, which carry out few calculations, but a lot of control of the interactions between their components and with their environment. The methods for specifying these two types of systems are different, since the transformational systems are specified using pre- and post-conditions, while reactive systems are specified declaratively (usually by temporal properties) and operationally, by a set of transitions between states, guarded by conditions of transition, called *guards*.

The reactive systems studied are said to be *parameterized*, because they consist of any (fixed) number of processes, and *uniform*, because all these processes are identical. The executions are assumed to respect the interleaving hypothesis, according to which at most one process evolves simultaneously with each event. The behavior of the system derives from the interactions of the processes with each other and the environment. These reactive systems are said to be *closed* when the hypotheses concerning the environment and the events likely to occur are taken into account in the modelling, for example using non-deterministic transitions. The studied properties are the simplest safety properties, which require that no execution of the system reaches certain states, called *critical*. A typical example, used in the course and in the first exercises, is the *mutual exclusion property*, which requires exclusive access to a shared resource between all processes. The safety of a system is established by first looking for an *invariant candidate*, which is a characterization of a subset of the states of the system excluding the critical states, then by formally demonstrating that this formula is not satisfied by any critical state and that it effectively constitutes an *(inductive) invariant* of the system, satisfied by all the initial states of the system and preserved by the action of each of its transitions. During homework, the study of Cubicle certificates is an opportunity to better assimilate this definition of an inductive invariant. Indeed, all these certificates contain an explicit formalization, in first-order logic, of the condition on the initial states and of the condition of preservation by any transition.

3.2   Cubicle

The second lesson of the course presents the Cubicle tool (https://cubicle.lri.fr/), its main features and its input language. Cubicle is an open source model checker resulting from the thesis work of A. Mebsout [10]. This work improves and implements techniques of model checking modulo theory and invariant synthesis [5–8]. Confidence in Cubicle results is enhanced by producing certificates which are separately verifiable with Why3.

Cubicle's input language makes it possible to define variables and arrays that model the data of a system and its processes, to specify the sets of initial and critical states, as well as the transitions whose activation is conditioned on the existence of an n-tuple of processes allowing a guard to be crossed. This language is documented in the thesis of A. Mebsout [10] and in the materials for a course given by S. Conchon in a school for young researchers [4]. The second lesson of the course *Specify and Verify* describes pedagogically and in detail the syntax and semantics of the fragment of this language useful for this course, illustrating them with simple examples. This exempts the students from having to consult external sources of documentation.

Cubicle is dedicated to the specification and verification of parameterized reactive systems of any size whose states are described by global variables and arrays. It allows the modeling of systems evolving in a discrete and non-deterministic way, under the action of guarded transitions. In order to favor the verification process, the internal subtleties of Cubicle are deliberately not detailed in the course, which only mentions that Cubicle uses a backward reachability algorithm, which goes back the transitions from critical states to try to reach the initial states. The discovered states, even if they are not critical, are *dangerous states* insofar as they are the starting point of at least one path leading to a critical state.

The course illustrates the modeling and verification process with Cubicle using a variant of the distributed mutual exclusion algorithm created and named "bakery" by L. Lamport [9]. This system models a bakery, in which customers obtain a numbered ticket that defines the order of waiting before placing an order with the baker. The customers, in any number, are the processes. In the original algorithm, the order of access to the baker is defined according to a lexicographic order relating to the ticket numbers and, in the event of a tie, the numbers identifying the customers. In the course variant, the ticket numbers issued to customers are distinct and are sufficient to determine the order of service. Therefore, the process that can access the critical section can always be known, if it exists.

In Cubicle, the states of the global system for "bakery" are defined by the following types and variables:

```
type status = WA | SE | AS
var Ticket : int
array CustomerStatus [proc] : status
array CustomerTicket [proc] : int
```

The `Ticket` variable, of integer type, stores the value of the last ticket delivered by the ticket dispenser. The `CustomerStatus` array stores the status of each customer, among the three states of the enumerated type `status`, described later. The `CustomerTicket` array stores the ticket value for each customer. For a customer without a ticket, this value is arbitrary.

All customers behave the same way, as follows. A customer in the requesting state `AS` (for ASking) can spontaneously obtain a unique numbered ticket. The number on this ticket is obtained by incrementing the counter `Ticket`. Once in possession of this ticket, this customer enters the waiting state `WA`. This action is formalized by the following Cubicle transition, parameterized by the process identifier `i`:

```
transition getTicket (i)
  requires { CustomerStatus [i] = AS } {
  CustomerStatus [i] := WA;
  Ticket := Ticket + 1;
  CustomerTicket [i] := Ticket + 1;
}
```

The access to the shared resource is formalized by the Cubicle transition `access` reproduced in Listing 1.1. If there is at least one customer waiting, then the baker serves the customer who has the smallest ticket (among the tickets of the waiting and served customers). The customer then switches to the served state SE.

Listing 1.1. Access to the served state, in Cubicle syntax

```
transition access (i)
  requires { CustomerStatus[i] = WA &&
    forall_other j.
      CustomerStatus[j] = WA || CustomerStatus[j] = SE
      => CustomerTicket[i] <= CustomerTicket[j] } {
  CustomerStatus[i] := SE;
}
```

Once served, a customer releases her/his place by switching to the AS state of customers likely to (re)request a ticket. This action is formalized by a simple Cubicle transition, named leave, which is not detailed here.

As specified in the following Cubicle clause, parameterized by the process identifier i, all customers are initially assumed to be ticket requesters (status AS) and to have a ticket numbered 0, which is also assumed to be the last ticket delivered by the ticket dispending machine. It may seem problematic that all customers initially have the same ticket number, but this is not an issue because no transition uses the ticket number of a requesting customer to determine the effects of a transition or to evaluate a guard. As the system evolves, the tickets owned by waiting and served customers are unique before being examined to decide who is the next served customer.

```
init (i) {
  CustomerStatus[i] = AS && CustomerTicket[i] = 0 &&
  Ticket = 0
}
```

Critical states are declared with one or more unsafe specifications. The following example expresses the existence of two distinct processes i and j in the critical state SE, which fails mutual exclusion.

```
unsafe (i j) {
  CustomerStatus[i] = SE && CustomerStatus[j] = SE && i <> j
}
```

When the modeling is finished, the execution of Cubicle in command line outputs a result concerning the safety of the system, reproduced in Fig. 1. In addition to the SAFE or UNSAFE verdict, Cubicle indicates sets of critical or dangerous states computed by the backward reachability algorithm. These are the *nodes* 1 to 8. A trace indicates a succession of transitions which makes it possible to reach a set of critical states from a set of dangerous states.

When the system is not safe, Cubicle returns an error trace indicating the transitions to follow to reach a critical state from an initial state. The student

```
guest@c33289fc8d66:~/data$ cubicle Bakery.cub
node 1: unsafe[1]
node 2: access(#2) -> unsafe[1]
node 3: access(#1) -> access(#2) -> unsafe[1]
node 4: getTicket(#2) -> access(#2) -> unsafe[1]
node 5: getTicket(#2) -> access(#1) -> access(#2) -> unsafe[1]
node 6: access(#1) -> getTicket(#2) -> access(#2) -> unsafe[1]
node 7: leave(#2) -> access(#1) -> getTicket(#2)
   -> access(#2) -> unsafe[1]
node 8: access(#2) -> leave(#2) -> access(#1)
   -> getTicket(#2) -> access(#2) -> unsafe[1]
...
The system is SAFE
```

Fig. 1. Result of a safety check with Cubicle.

can check this trace by reconstructing the evolution of the system manually. Cubicle does not offer a tool to automate this trace analysis.

Cubicle also makes it possible to generate a certificate of this proof, in the input language of the Why3 platform. A certificate contains a logical description of the verified system, a candidate invariant and goals to prove, which formalize that the candidate invariant excludes critical states, and that it is indeed an invariant of the system, satisfied by all initial states and preserved by all transitions of this system.

Implementing verification with Cubicle is therefore straightforward. If it is possible to write the specifications of a system respecting the Cubicle input language syntax, then the system is verifiable with Cubicle. Otherwise, you have to turn to other tools, such as the Why3 platform, which is more generic.

3.3   Why3 Platform

The Why3 platform enables proofs of propositional or first-order logic formulas, or of conformity between imperative or functional programs and their logical specification. Its language, called WhyML, is more general and more complex than that of Cubicle. Its use allows students to discover and implement the notions of contract, loop invariant, etc., and to develop several skills. First, they learn to solve decision problems with Why3, formalizing them as lemmas or goals. For simplicity, the course is limited to the case where these logical formulas concern variables of predefined Boolean or integer type, or of enumerated type defined by the user, and/or of array type storing data of these types. Then, the students learn to specify in WhyML a simple imperative program contract, manipulating the same data types, and to annotate its loops, until making the verification of this contract automatic with the Why3 platform. In addition, students learn to read and modify certificates generated by Cubicle. Finally, students learn to specify reactive systems directly in WhyML, possibly beyond the expressiveness limits of Cubicle, and to verify their safety with Why3, as detailed in Sect. 3.4.

3.4 Specification and Verification of Reactive Systems only with Why3

Through its input language and its internal mechanisms for invariant strengthening and WhyML certificate generation, Cubicle greatly facilitates the specification and verification of parameterized reactive systems. However, it is a black box, applicable only to systems whose sets of critical states can be described by formulas, called *cubes*, whose syntax is limited to conjunctions of literals existentially quantified by identifiers of distinct processes. How to check a property of a system that cannot be specified in this way?

So that this course remains introductory, it does not provide a general answer to this question, which would be too technical, but provides intuitions based on examples. In the first place, some advanced exercises, or even some questions of an exam subject, may propose to the students to modify a certificate generated by Cubicle, for example to add a property to check. However, this first approach has a major limitation, inherent to Cubicle's translation of transitions into logical relations: the user cannot execute these specifications. But WhyML is also a programming language, of which a fragment is directly executable, and a larger fragment is executable by extraction in OCaml. On the example of the bakery algorithm, the rest of this section presents a way to describe reactive systems by WhyML programs, and then discusses the issue of their verification.

The following code proposes a way to describe the states of the bakery system in WhyML, for a number of customers fixed by the constant n, which must be a positive integer (condition Pos, required because the WhyML type int corresponds to relative mathematical integers).

```
type status = WA | AS | SE
val constant n : int
axiom Pos : n > 0
type sys = {
  mutable ticket : int;
  customerStatus : array status;
  customerTicket : array int
} invariant {
  invariant_candidate ticket customerStatus customerTicket
} by {
  ticket = 0;
  customerStatus = make n AS;
  customerTicket = make n 0
}
```

Cubicle variables and arrays are grouped here as fields of a record of type sys. These fields are all *mutable*, i.e. modifiable in place, either because they are declared with the mutable keyword, or because they are arrays, always mutable in WhyML. The type sys is defined with a *type invariant*, which imposes conditions on its fields. These conditions are hidden here behind the predicate invariant_candidate. Some technical conditions require the ticket numbers to be non-negative integers, and the two arrays customerStatus and customerTicket to be of the same size n. The other conditions characterize any set of supposedly safe states, with respect to given critical states.

The WhyML language requires that any type with an invariant be declared with a by clause that justifies that the type is not empty, by describing an inhabitant of it. Here, we suggest as inhabitant the initial state of the bakery system, with the provided WhyML function make, which constructs an array whose elements all have the same value. When verifying a type declaration with an invariant, Why3 tries to prove that the state described by its by clause satisfies the type invariant described by its invariant clause. By this design choice, without anything else to write, we obtain the verification condition that the invariant candidate is true in the initial state of the bakery system.

In order to formalize in WhyML the non-deterministic choice of a process which satisfies the guard of a transition, the abstract function any_int_where, specified by the following code, is provided to students. It models the non-deterministic choice of an integer i satisfying the condition (p s), for any given executable predicate p and any inhabitant s of any type 'a. (In WhyML, as in many functional languages, the application of the function f to the argument x is denoted f x, instead of f(x)).

```
val any_int_where (s: 'a) (p: 'a -> int -> bool) : int
   requires { exists i. p s i } ensures { p s result }
```

As illustrated by an example below, in the WhyML code of a transition parameterized by a single process, the type 'a will be the type of the states, the variable s will be a state and the predicate p will be an executable version of the guard of this transition. The function any_int_where only exists if there exists at least one integer i satisfying the condition p s. This is required by the precondition (exists i. p s i). Under this condition, the postcondition (p s result) expresses that the function returns such an integer, represented in this expression by the reserved word result of the WhyML language.

In the presence of a declaration with the keyword val, Why3 admits that the declared object – here, a function – exists, without requiring or providing any justification for this existence. However, if such a function cannot exist, assuming the contrary would make the logic inconsistent. In order to show that this is not the case, we can provide the following implementation of this function.

```
let any_int_where (s: 'a) (p: 'a -> int -> bool) : int
   requires { exists i. p s i } ensures { p s result }
= any int ensures { p s result }
```

The expression any ... ensures ... non-deterministically evaluates to a value that satisfies the logical formula after the keyword ensures, when this value is assigned to the dummy variable result. For this expression, Why3 always produces a verification condition that this logical formula is indeed satisfiable. Due to the precondition (requires clause), this condition of existence of the function any_int_where is easily discharged by the Why3 platform.

Then, each transition of the system is implemented as a WhyML function, as in the example reproduced in Listing 1.2. The guard of this transition is that there is a customer in the AS state. This guard is defined by the predicate

Listing 1.2. Transition of a customer obtaining a ticket, in WhyML

```
let predicate p_getTicket (s: sys) (c: int)
= 0 <= c < length s.customerStatus &&
  s.customerStatus[c] = AS
let getTicket (s: sys) : sys
  requires { exists i. p_getTicket s i }
= let c = any_int_where s p_getTicket in
  s.customerStatus[c] <- WA;
  s.customerTicket[c] <- s.ticket + 1;
  s.ticket <- s.ticket + 1;
  s
```

p_getTicket, which is also a Boolean function, thanks to the keyword `let` which allows to use it in programs. A technical point here is that equality (=) is defined in WhyML as a logical predicate for any type, but does not exist by default as a Boolean function on user-defined types, such as `status`. For executability, a Boolean equality on the type `status` must be implemented, for example by the following code.

```
let (=) (x y: status) : bool ensures { result <-> x = y }
= match x with
  | WA -> match y with WA -> True | _ -> False end
  | AS -> match y with AS -> True | _ -> False end
  | SE -> match y with SE -> True | _ -> False end
  end
```

A second technical point concerns guards that include a quantified condition on processes, such as the guard of the `access` transition. In order for the corresponding predicate to be executable, students are provided with the following Boolean function for universal quantification over a bounded integer interval.

```
let predicate forAll (s:'a) (p:'a->int->bool) (l u:int)
  ensures { result <-> forall i. l <= i <= u -> p s i }
= for j = l to u do
    invariant { forall i. l <= i < j -> p s i }
    if not (p s j) then return False
  done;
  True
```

The `any_int_where` function is similar to the ANY …WHERE …THEN …END construct of the B language, and to the `any` keyword of WhyML, whose execution in a program realizes a non-deterministic choice of a typed value satisfying a given condition. Instead of using this instruction, we have preferred to provide and suggest to use the `any_int_where` function, in order to facilitate various implementations of non-determinism. Indeed, it is simpler to associate an implementation to a single function than to have to replace each `any` expression by an implementation.

Each transition is a WhyML function parameterized by the complete system (named s in this example, of type `sys`) and which returns the new state of this system, of the same type `sys`. When the Why3 tool verifies such a function, it tries to show that the returned state satisfies the invariant of type `sys`, under the hypothesis that its parameter `s` satisfies this invariant. This verification condition is exactly the second condition for this type invariant to be an invariant of the transition system, without anything else to specify about it. Without this use of a Why3 type with an invariant, it would have been necessary to repeat the invariant candidate as a precondition and postcondition of each transition, which is more cumbersome and presents a risk of error by omission.

When the system is fully programmed, the student verifies its safety by calling the SMT solvers available through the Why3 platform. If the proof of safety succeeds, the exercise comes to an end. Otherwise, the student has to verify her/his code and specifications, to reinforce the invariant candidate or to question the capacities of the SMT solvers to discharge the verification conditions. As these last two tasks are difficult, the exercises are accompanied by advice.

A *reinforcement* of an invariant candidate is any additional formula that approximates more accurately the reachable states, and whose conjunction with the invariant candidate is expected to form an inductive invariant. As first invariant candidate, it is natural to choose the negation of a characteristic predicate for critical states. For any mutual exclusion algorithm, such as the bakery system, this invariant is the mutual exclusion property, saying that no two distinct processes simultaneously use the shared resource. For the bakery system, this property of simultaneous non-existence of two served customers is not an inductive invariant. Indeed, this property holds for instance in the state with two customers, one served and the other one waiting and holding a smaller ticket. From this state, the second client can be served by the transition `access` (whose code is reproduced on Listing 1.1), leading to the critical state. Fortunately, the combination of the other transitions and the initial states makes these dangerous states unreachable. The Cubicle output

```
node 2: access(#2) -> unsafe[1]
```

on the third line of Fig. 1 means that Cubicle identifies these dangerous states and choose their negation as first invariant reinforcement. Nodes 3 to 8 reproduced in Fig. 1 similarly correspond to sets of dangerous states found by Cubicle, and used by Cubicle to produce other reinforcements included in the certificate. Their conjunction with the initial invariant candidate forms an inductive invariant.

When we choose as reinforcements the invariants synthesized by Cubicle, and provided to us in the certificate it generates, and we encode all the guards with `let predicate`, then Why3 automatically proves that the resulting invariant is preserved by the `getTicket` and `leave` transitions, but not by the `access` transition. This last proof becomes automatic if we separately define the guard by a predicate and a Boolean function, along the pattern reproduced in Listing 1.3.

Figure 2 illustrates the proof results when the system is specified in this way. The verification condition `sys'vc` is the condition that the initial states satisfy the invariant, while the verification conditions associated with the transitions deal with the preservation of the invariant.

Listing 1.3. Pattern for the access transition, in WhyML

```
predicate p_access (s: sys) (i: int) = ...
let g_access (s: sys) (i: int) : bool
  ensures { result <-> p_access s i } = ...
let access (s : sys) : sys
  requires { exists i. p_access s i }
= let c = any_int_where s g_access in
  s.customerStatus[c] <- SE;
  s
```

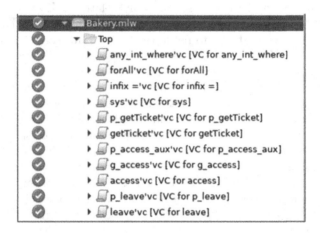

Fig. 2. Verification of a bakery system with the Why3 platform

Generally speaking, finding reinforcements is all but an easy task, beyond the expected level of the students at the end of this course. If the exercise could be handled with Cubicle, students are asked to adapt the invariants that Cubicle wrote in its proof certificate. Otherwise, some exercise questions suggest reinforcements in natural language, which the student only has to formalize in WhyML.

In conclusion, while the process of verification of parameterized reactive systems with Cubicle is relatively simple, this process with the Why3 platform is more complex, since reinforcements may be required to complete the proofs. Therefore, exercises and homework are necessary to facilitate its learning by students.

4   Virtual Machine for Labs

Each year, the first exercise of the first homework guides the students to build their lab environment, in the form of a Docker (https://www.docker.com) container. At the end of this exercise, the students have a virtual machine (called a *Docker container*) in which they can run Cubicle (version 1.1.2) and Why3 (version `1.4.0`) software command line, but also the graphical interface of Why3. The virtualization technology Docker is available for Linux, macOS, and Windows. The Docker container being the same virtual machine for all students, it limits installation problems and allows tutors to reproduce the actions of the students identically, and thus better help them to use the tools. Using this container allows students to do homework with their personal computer without disrupting their work habits. A working directory is shared between the host machine and the container. Files for Cubicle and the Why3 platform stored in this directory can be accessed and edited from the host machine or in the running container.

The Docker image of this working environment is formally described in a `Dockerfile` provided to students. It expands the Docker image `registry.gitlab.inria.fr/why3/why3:1.4.0` distributed by Inria, which contains the Why3 platform (version `1.4.0`) and the three CVC4 SMT solvers (version `1.7`), Alt-Ergo (version `2.0.0`) and Z3 (version `4.8.4`). The `Dockerfile` complements this image with Opam (*OCaml package manager*, https://opam. ocaml.org) and an installation of the Cubicle software with Opam. For Windows users, a VcXsrv server (https://sourceforge.net/projects/vcxsrv) is used as an X server, for the graphical interface of the Why3 platform. The creation and starting of the working environment are made easier thanks to provided scripts for Linux and macOS, and `.bat` batch files for Windows.

No integrated development environment (IDE) is suggested to edit input files for Cubicle. Any text editor is suitable, even without syntax highlighting, since Cubicle's input language is very readable and the requested models are achievable in a few dozen of lines.

The creation of the lab environment being an essential step, the students are invited to communicate on the mutual aid forum the difficulties they encountered. The exercise represents 8 points out of 20 in the grade for the first homework. Its evaluation is carried out on the basis of a report on the application of the homework subject instructions and the student's participation in the mutual aid forum on this activity.

5   Evaluations

We first describe how students are evaluated, and then how the teaching unit is perceived by its students.

5.1   Student Assessment

The exercises are not graded. Homeworks being works done without time limit and without supervision, their grade are not taken into account in student assessment. These grades, however, have an indicative value: they quantify the quality of the content of the returned assignments and inform students of the level expected at the exam.

By using verification tools during their works at home, the students see themselves whether their answers are correct. These verification tools could also be used to automatically generate a grade based on the number of successful verifications. However, such an automated grading tool is of little interest, for several reasons. First, automated verification for complete specifications being rare, it can only be one element of evaluation among others, with a small coefficient. The other evaluation criteria relate to the understanding of the concepts, which requires a human analysis producing personalized written recommendations. Finally, the number of students returning homework being low (from 10 to 20), automated grading would not represent a significant time saving for teachers.

The students are evaluated on 20 points, during a written final exam (on paper), supervised and lasting 2 h. The examination must be carried out without using electronic devices and without consulting documents other than the examination subject sheets. The fragments of the Cubicle and Why3 language syntaxes useful for this subject are recalled in an appendix at the end of the subject. Thus, the only noticeable difference between the exam requirements and what the students are trained to do during the semester is that they cannot run the Cubicle and Why3 tools during the exam to detect errors in the codes they are writing on their exam copy. This limitation encourages students to produce (almost-)correct code more rationally, without resorting too much to the empirical trial-and-error cycle. The negative effect is minimized by an evaluation that ignores minor syntax errors. Students are informed of the exam conditions at the start of the semester, and can practice them thanks to annals provided with their answers. The *Specify and Verify* module counts for 3 ECTS credits (*European Credit Transfer System*) among 30 for one semester of the master.

5.2   Course Assessment

Once the course, the exercises, the homework and the final exam have been practiced by the students, an evaluation is carried out to ensure that the module finds a favorable reception. No automatic evaluation system is used, because the numbers of students and hours devoted to this teaching are small. The evaluation of the course is carried out directly by questioning the students, either during an in-person review meeting, or through a digital forum during the COVID-19 pandemic. It appears, for example, that the speed and efficiency of the use of the Cubicle *model checker* is appreciated by the students. Difficulties encountered by the students during exercises and homework, expressed as questions in the forum, are taken into account as they arise, in the form of additional explanations or course modifications. Final exam results also help determine course improvements for the following year.

6   Discussion

The prerequisites to follow the formal verification process taught in the *Specify and Verify* module are few. The students have generally already practiced the modeling of systems with states evolving under the effect of transitions. Reminders concerning discrete event systems and first-order logic, as well as short introductions to the Cubicle and Why3 tools, are sufficient to facilitate the acquisition of new knowledge and know-how.

If reading and writing formulas in first-order logic is an uncommon practice in the context of students' previous programming activities, writing specifications in logic remains accessible to them, since the guards of the transitions to be formalized correspond to simple logical formulas. The difficulty in proving certain parameterized reactive systems lies above all in the design of an invariant approximating the non-dangerous states, given the specifications. This difficulty is reduced thanks to the automated invariant reinforcement mechanism implemented in Cubicle.

The formal approach to specification and verification of parameterized reactive systems provides students with a concrete example of *symbolic model checking* based on the decision procedures implemented in SMT (Satisfiability Modulo Theory) solvers. However, no temporal logic is taught to specify the properties to be checked, since the only property dealt with is safety, which is reduced to a state reachability analysis. The usefulness of this verification by exploration (of sets) of states (model checking), and of its automation, are highlighted by offering students examples of systems whose reachable states are difficult to predict intuitively, while their transitions are simple to define. In particular, the search for reinforcements of invariants is not very intuitive, which illustrates the difficulty of predicting the states reached by the system.

The teacher who designs the exercises must ensure that safety is not trivially ensured by the system, but is really a property emerging from the system specifications. Otherwise, checking it becomes obvious and its formal verification loses its interest. For instance, mutual exclusion does not constitute an emergent property of a system whose transitions would explicitly check (in their guard) that no other process owns the critical resource.

When an invariant has been proven, it is possible to continue the study of the system to verify other complementary properties. The proven invariant is completed with an additional property and a new verification of the preservation of the invariant is carried out. If the modified invariant remains preserved, then the additional property is verified for all the states reached by the system.

7   Conclusion

The *Specify and Verify* module is built around the use of the model checker Cubicle and the Why3 platform. These tools have proven to be suitable for teaching the verification of the safety of simple parameterized reactive systems. This practical approach to verification is not very demanding in terms of prior knowledge and formalization skills. It provides students with a concrete example of model checking. However, when a system is safe, but it is necessary to carry out reinforcements oneself to find an invariant, the effectiveness of verification with the Why3 platform is greatly reduced.

Although it is possible to use the invariant reinforcement technique applied by Cubicle in a black box, it would be interesting to add in this module a lesson on invariant reinforcement (semi-)algorithms, such as those of theses of J.-F. Couchot [5] and A. Mebsout [10], also to show how to formalize them in WhyML and to verify some of their properties with Why3. This would add to the module a complementary aspect of formal semantics, in particular about non-determinism.

Acknowledgments. This project is supported by the EIPHI Graduate School (contract ANR-17-EURE-0002). We thank the three anonymous reviewers for their comments and suggestions that helped us improve our original manuscript.

References

1. Blazy, S.: Teaching deductive verification in Why3 to undergraduate students. In: Dongol, B., Petre, L., Smith, G. (eds.) FMTea 2019. LNCS, vol. 11758, pp. 52–66. Springer, Cham (2019). https://doi.org/10.1007/978-3-030-32441-4_4
2. Bobot, F., Filliâtre, J.-C., Marché, C., Melquiond, G., Paskevich, A.: The Why3 Platform Release 1.5.1 (2022). https://why3.lri.fr/manual.pdf
3. Conchon, S., Goel, A., Krstić, S., Mebsout, A., Zaïdi, F.: Cubicle: a parallel SMT-based model checker for parameterized systems. In: Madhusudan, P., Seshia, S.A. (eds.) CAV 2012. LNCS, vol. 7358, pp. 718–724. Springer, Heidelberg (2012). https://doi.org/10.1007/978-3-642-31424-7_55
4. Conchon, S.: Model checking, Part 1: Model checking modulo theories (MCMT) (2015). cours de l'EJCP (École des Jeunes Chercheurs en Programmation). https://www.lri.fr/~conchon/EJCP/ejcp-mcmt.pdf
5. Couchot, J.-F.: Vérification d'invariants de systèmes paramétrés par superposition. Ph.D. thesis, Laboratoire d'Informatique de l'Université de Franche-Comté, Besançon, France (2006)
6. Couchot, J.-F., Giorgetti, A., Kosmatov, N.: A uniform deductive approach for parameterized protocol safety. In: Proceedings of the 20th International Conference on Automated Software Engineering, ASE 2005, pp. 364–367. IEEE (2005)
7. Ghilardi, S., Ranise, S.: Goal-directed invariant synthesis for model checking modulo theories. In: Giese, M., Waaler, A. (eds.) TABLEAUX 2009. LNCS (LNAI), vol. 5607, pp. 173–188. Springer, Heidelberg (2009). https://doi.org/10.1007/978-3-642-02716-1_14
8. Ghilardi, S., Ranise, S.: Backward reachability of array-based systems by SMT solving: termination and invariant synthesis. Logical Methods Comput. Sci. **6**(4) (2010)
9. Lamport, L.: A new solution of Dijkstra's concurrent programming problem. Commun. ACM **17**(8), 453–455 (1974). https://doi.org/10.1145/361082.361093
10. Mebsout, A.: Invariants inference for model checking of parameterized systems. Ph.D. thesis, Université Paris Sud - Paris XI (2014). https://tel.archives-ouvertes.fr/tel-01073980

Model Checking Concurrent Programs for Autograding in pseuCo Book

Felix Freiberger[(✉)] [ID]

Computer Science – Saarland Informatics Campus,
Saarland University, Saarbrücken, Germany
`freiberger@depend.uni-saarland.de`

Abstract. With concurrent systems being prevalent in our modern world, concurrent programming is now a cornerstone of most computer science curricula. A wealth of platforms and tools are available for assisting students in learning concepts of concurrency. Among these is pseuCo, a light-weight programming language featuring both message passing and shared memory concurrency. It is supported by pseuCo Book, an interactive textbook, focusing on the theoretical foundations of pseuCo, concurrency theory. In this paper, we extend pseuCo Book with a chapter on *Programming with pseuCo*. At its core is a custom verification system, based on pseuCo's Petri net semantics, enabling practical programming exercises to offer fast in-browser model checking that can validate the program's internal use of concurrency features and provide comprehensive debugging features if a fault is detected.

Keywords: Verification · Model checking · Autograding ·
Concurrency · Education · Colored Petri nets · Programming

1 Introduction

Since its infancy, computer science education has been a task of growing importance – and difficulty. The ever-growing use of concurrency, with multi-core CPUs now being prevalent even in embedded devices, has certainly furthered that trend. Today, instruction in practical concurrent programming, and the theoretical underpinning needed to fully understand it, is an essential component of a complete computer science curriculum.

Instruction in practical computer science often involves having students write programs. This has created interest in *autograding*, technologies to automatically check solutions for mistakes and provide feedback or assign a grade [10,11]. While this may seem like an excellent use case for verification techniques, in practice, autograding is often based on testing. This approach is powerful in many cases, e.g. even allowing autograding of full Android applications [2], however, testing-based autograding is particularly challenging in concurrency-related exercises, as concurrent programs are usually nondeterministic and may have bugs that are hard (or even impossible) to detect in a test environment [3].

C. Dubois and P. San Pietro (Eds.): FMTea 2023, LNCS 13962, pp. 51–65, 2023.
https://doi.org/10.1007/978-3-031-27534-0_4

Moreover, independent of whether testing or stronger verification-based techniques are used in autograding, they are often used to verify a program's externally visible behavior. While this often is sufficient, in many cases, a deeper look into a program's internals is required to ensure students have solved an exercise in the intended way. This is a typical requirement in introductory-level concurrency exercises which are often exploitable with easy non-concurrent solutions that have the same externally visible behavior than the concurrent program students were intended to write.

In this paper, we present a verification-based approach for autograding introductory-level concurrent programming exercises. Our approach is built around pseuCo [1], an academic programming language designed to teach students the basics of message passing and shared memory concurrency. We use pseuCo's Petri-net-based semantics [5] to gain insight into the semantics of the pseuCo program under analysis, enabling verification of properties about the use of concurrency-related features. For example, this allows us to not only verify the output of a program, but confirm that it uses channel-based communication between agents in a predetermined way, or that it is free of data races.

Our verification system is deeply integrated into pseuCo Book [4], a web-based interactive textbook focused around teaching concurrency theory and practice. It backs the interactive exercises in a new *Programming* chapter of pseuCo Book that guides students through their first contact with message passing and shared memory features, ensuring that the exercises only accept programs using these constructs correctly and as intended. When a student's program fails verification, we use the pseuCo Debugger, a Petri-net-backed debugging tool for pseuCo programs [5] originally developed for the web IDE pseuCo.com [1], to display failure traces to students in a way that is easily understandable without any knowledge about verification technologies or Petri nets.

Structure of this Paper. The remainder of this paper is structured as follows. Section 2 formalizes the properties that are to be analyzed. Section 3 documents the implementation of the model checker as part of pseuCo Book. Section 4 describes the new Programming chapter of pseuCo Book that is supported by this technology. Finally, Sect. 5 concludes this paper.

2   pseuCo Program Properties

2.1   Motivating Example: Message-Passing-Based Termination

In this section, we'll look at an example exercise from pseuCo Book.

PseuCo's message passing features allow the programmer to use synchronous and asynchronous channels to transfer primitive values between agents. An example of this – typically one of the first pseuCo program students see – is printed in Listing 1.1. This program computes the value of (3!)! using a `factorial` agent that computes the factorial of any number it receives on a synchronous (handshaking) channel, then sends it back on the same channel.

Listing 1.1. A simple message passing pseuCo program

```
1  void factorial(intchan c) {
2      int z, j, n;
3      while (true) {
4          z = <? c; // receive input
5
6          n = 1;
7          for (j = z; j > 0; j--) {
8              n = n*j;
9          }
10
11          c <! n; // send result
12      };
13  }
14
15  mainAgent {
16      intchan cc;
17      agent a = start(factorial(cc));
18      cc <! 3;
19      int mid = <? cc;
20      println("3!␣evaluates␣to␣" + mid + ".");
21      cc <! mid;
22      println("(3!)!␣evaluates␣to␣" + (<? cc) + ".");
23  }
```

While receiving messages from a specific channel is relatively straightforward, in some cases, a programmer may need to set up an agent to react to multiple possible message passing actions, e.g. incoming messages on two different channels. Doing so requires a dedicated language construct, which pseuCo borrowed from Go: the **select case** statement.

To teach students how to use this statement, pseuCo Book contains an exercise asking students to modify the program from Listing 1.1 such that the **factorial** agent terminates after it is no longer needed by the main agent.

There are two apparent methods to do so:

1. reserve a special value, like -1, that triggers the agent's termination, or
2. add a second, dedicated channel for termination requests.

The first method does introduce a corner case, so the exercise asks students to add termination cleanly, using the new **select case** statement to add a Boolean-typed control channel.

What does an autograder need to check when validating a solution to this exercise? Adding termination does *not* actually change the externally visible behavior of the program[1]. But even if the autograder was able to determine whether all agents have terminated at the end of execution, this would not actually test whether students have implemented the exercise in the intended way. For example, students could replace the **while (true)** loop with a hardcoded **for (int i = 0; i < 2; i++)** loop to cause the agent to terminate after two

---

[1] The pseuCo semantics does not allow externally distinguishing between termination and a deadlock.

iterations which would work in this specific example, but not in general (and does not demonstrate knowledge of how to use the `select case` statement).

While some of these pitfalls can be overcome by testing the students' submission in multiple, slightly different contexts, a more thorough solution is for the autograder to also inspect the program's internals, checking that every possible execution of the program completes these steps in order:

1. start an agent (agent 1) from the main agent (agent 0)
2. use synchronous communication to send the value 3 from agent 0 to agent 1
3. use synchronous communication to send the value 6 from agent 1 to agent 0
4. have agent 0 print `"3!␣evaluates␣to␣6."`
5. use synchronous communication to send the value 6 from agent 0 to agent 1, with agent 1 being in a `select case` statement with 2 cases for receiving values
6. use synchronous communication to send the value 720 from agent 1 to agent 0
7. and then, in any order
   - print `"(3!)!␣evaluates␣to␣720."` from agent 0
   - complete these steps in order:
     (a) use a synchronous channel to send a Boolean from agent 0 to agent 1, with agent 1 being in a `select case` statement with 2 cases for receiving values
     (b) terminate agent 1

Indeed, this is the approach we will follow. The following sections formalize this type of property.

2.2   pseuCo Verification Formalities

Let *pseuCo* be the set of pseuCo programs. We define colored Petri nets following Jensen [8], i.e. as a tuple $CPN = (\Sigma, P, T, A, N, C, G, E, I)$ where Σ is the set of color sets, P the sets of places, T the set of transitions, A the set or arcs, N the node function, C the color function, G the guard function, E the arc expression function and I the initialization function. For clarity, we refer to the steps of a colored Petri net's execution as *firings*.

The pseuCo compiler [5] translates every valid pseuCo program $p \in pseuCo$ into a colored Petri net *CPN* and *labels* : $P \cup T \mapsto 2^{\mathcal{L}}$, a *pseuCo label function* that assigns sets of labels to both places and transitions. The set of labels $\mathcal{L}$, not described in full detail here, contains labels that describe the role of places and transitions in pseuCo terminology. For example, a place could be labeled (global-variable, "x") to indicate that it holds the value of a global variable named x, or a transition could be labeled (send-async) to indicate that it handles sending a message to an asynchronous channel (i.e. writing the value to its buffer).

To formalize our properties, we use LTL [9]. We assume a set AP of atomic propositions, deferring details to the next section. Skipping details for brevity, we assume a mapping from firings of the Petri net to subsets of atomic propositions.

The properties that are relevant for autograding pseuCo exercises can then be expressed as *LTL formulas*, i.e. terms φ with

$$\varphi ::= \neg\varphi \mid \varphi \vee \varphi \mid \varphi \wedge \varphi \mid \varphi \rightarrow \varphi \mid \bigcirc\varphi \mid \Box\varphi \mid \Diamond\varphi \mid \varphi\,\mathcal{U}\,\varphi \mid ap \qquad (1)$$

and $ap \in AP$.

Atomic Propositions. Using the pseuCo label function (and knowledge of the internals of the pseuCo-to-CPN compiler), a Petri net firing can be analyzed to determine which pseuCo action the firing represents. This allows us to define and recognize atomic propositions that describe whether a firing

- prints a specific value,
- has specific agents participate in that firing (identified by their IDs),
- has a (single) participating agent with a specific expected recursion depth,
- represents a synchronous message passing transaction (handshaking), with a specified value,
- represents writing to or reading from an asynchronous (buffered) channel, with a specified value,
- originates from a `select case` statement (with a certain number of branches),
- starts or terminates an agent,
- represents a procedure call (with specific arguments),
- initializes, locks, or unlocks a lock,
- reads or writes a global variable (with a specific name and value), and
- reads or writes a structure field (with a specific name and value).

These atomic propositions can then be used by exercise designers to describe the intended behavior of a pseuCo program without insight into the specifics of the pseuCo-to-CPN compiler.

Step Checklists. Generally, the full flexibility of LTL is not needed to express the properties used for autograding pseuCo exercises. To simplify the process of specifying these properties – and allow representing the property, and its current state, more easily to a user – we introduce a simplified syntax for these properties, called *step checklists*. The set of step checklists S is defined as

$$s_1, \ldots, s_n \ni S ::= \text{Step}\,(v_1) \mid \text{Sequence}\,(s_1, \ldots, s_n) \mid \text{Parallel}\,(s_1, \ldots, s_n) \quad (2)$$

$$v_1, \ldots, v_n \ni V ::= v_1 \wedge v_2 \mid v_1 \vee v_2 \mid \neg v_1 \mid ap \quad (3)$$

with $ap \in AP$. Conceptually, a step checklist is a list of steps a pseuCo program has to complete – in a fixed order, in arbitrary order, or in arbitrarily nested fixed-order and free-order blocks.

Together with the atomic propositions described previously, step checklists allow a compact representation of properties. For example, the first message passing exercise in pseuCo Book uses the atomic propositions *startAgent* that holds when an agent is started, *agents(x)* that holds when the set of agents participating in a step is exactly x, *handshake(v)* that holds when value v is passed by handshaking, and *print(v)* that holds when value v is printed:

$$\text{Sequence} \begin{pmatrix} \text{Step}\,(startAgent)\,, \\ \text{Step}\,(agents(\{0,1\}) \wedge handshake(\texttt{"World"}))\,, \\ \text{Step}\,(agents(\{1\}) \wedge print(\texttt{"Hello,\textvisiblespace World!"})) \end{pmatrix} \quad (4)$$

This step checklist ensures the main agent starts an agent, passes `"World"` to it, after which that agent prints a greeting.

A step checklist s can easily be converted into an LTL property $[\![s]\!]$:

$$ltl(\text{Step}\,(v)) := v \tag{5}$$

$$ltl(\text{Sequence}\,(s_1, s_2, \ldots, s_n)) := ltl(s_1) \wedge \bigcirc \diamond (ltl(s_2) \wedge \bigcirc \diamond (\ldots ltl(s_n))) \tag{6}$$

$$ltl(\text{Parallel}\,(s_1, \ldots, s_n)) := (\diamond ltl(s_1)) \wedge \cdots \wedge (\diamond ltl(s_n)) \tag{7}$$

$$[\![s]\!] := \diamond ltl(s) \tag{8}$$

3 Verification

LTL formulas can be model checked efficiently by conversion to a Büchi automaton [6,7]. Here, we follow the same approach, with some optimizations and extensions specific to our use case.

3.1 Implementation and Integration into pseuCo Book

For use in pseuCo Book's programming exercises, the verification system has been implemented in JavaScript, based on the pseuCo-to-CPN compiler and the `colored-petri-nets` JavaScript library [5] for pseuCo Semantics. The step checklist created by the exercise designer is converted directly to an automaton, skipping the intermediate LTL step for efficiency. An exhaustive search of the cross product of this automaton and the reachability graph of the Petri net is then performed. For efficiency, the atomic propositions are not precomputed, but dynamically evaluated during search. Verification starts only on demand, when the user explicitly "submits" their program. All computation is done locally in the user's browser, allowing offline use, without relying on a centralized service. Using the Web Worker API, all computation is performed in a background thread, ensuring the UI stays reactive and verification can be cancelled if needed.

This verification technology backs the interactive exercise in pseuCo Book's new *Programming with pseuCo* chapter. It is not otherwise accessible to the user – notably, users cannot input new specifications.

When verification succeeds, the corresponding exercise is marked as solved. (Students can refine or re-do their solution and run verification again, but the exercise will continue to be marked as solved).

When verification fails, a failure trace is generated – a sequence of Petri net firings. Students are then presented with an error message stating that an execution of their program failed to meed the specification, as demonstrated in Fig. 1. To generate this error message, in the implementation, atomic propositions are associated with a human-readable description of the behavior they are looking for, which are then used to assemble the final error message, for example:

The program has terminated or deadlocked. It was expected to
send "Hello" on an asynchronous channel from the main agent.

In addition, users are given the option to inspect the failure trace. This trace by itself is not suitable to show to users of pseuCo Book as they are not expected to know Petri nets semantics nor the details of the pseuCo-to-CPN translation.

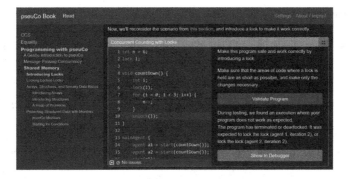

Fig. 1. Screenshot of a programming exercise in pseuCo Book containing a fault

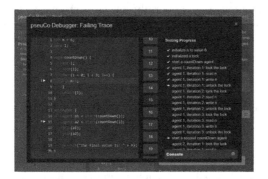

Fig. 2. Screenshot of a failing trace as shown in the debugger view

Fortunately, pseuCo.com, the web IDE for pseuCo, contains the *pseuCo Debugger* [5], a debugger-like interface that allows "executing" a pseuCo program like in a traditional IDE while maintaining full control over all possible executions allowed by the language specification. PseuCo Debugger is also built on top of the pseuCo-to-CPN toolchain – technically, it is a tool to explore the reachability graph of a colored Petri net, but it uses the labelling function *labels*[2], plus knowledge of the internals of the pseuCo-to-CPN compilation, to convert the marking into pseuCo terminology, fully hiding the Petri net and instead showing a debugger-style interface.

This allows us to use the existing UI of pseuCo Debugger to visualize failing traces. When verification fails, users are given an option to invoke the debugger. This launches a slightly modified version of pseuCo Debugger, shown in Fig. 2, with the following differences from the original version:

– The trace is fixed: The debugger opens with the full trace already pre-selected, and users have no option to view or change nondeterministic choices – they can only navigate forwards or backwards in time.

---

[2] The labelling function used by pseuCo Debugger is identical to the one used in the verification system.

– For every step of the trace, the debugger also shows the *verification state*: The step checklist is converted into a flat list, presented within the debugger interface like a todo list. (This does not allow users to see which steps are sequential and which are in parallel order, but a checkmark and an arrow will highlight items that are completed or currently due, respectively).

This allows the user to not only see the full sequence of actions their program took, presented in the style of a traditional high-level debugger, but also shows them how their program progresses (or fails to progress) through the step checklist, helping them discover where their program deviated from the specification.

3.2   Advanced Validation Features

While pseuCo Book's exercises generally use step checklists as described above, some exercises need additional expressivity for their properties, or additional features that do not fit into the theoretical framework described previously.

The following sections give an overview of these extensions as implemented in pseuCo Book.

Fail Fast and Cycle Detection. As described previously, step checklists always correspond to LTL formulas starting with ◇, and thus can only be violated by pseuCo programs terminating (or diverging) without having satisfied the requirements. While this is sufficient to express most relevant properties for the exercises in pseuCo Book, such properties often yield unhelpful error messages.

For example, consider a problem statement asking students to compute 3!, then printing it. Assume a student solution contains a mistake that causes the program to compute and print an incorrect solution. Then, the error message generated from the underlying LTL formula will complain that the program terminated without printing 6, which mosts students will find less helpful then being informed that their program printed an incorrect value.

To improve this, in the implementation, step checklists can also *ban* groups of atomic proposition. If such an atomic proposition is encountered while the corresponding step is active, verification is immediately terminated, and a custom error message is returned as the verification result. This can then be used by the exercise designer to explicitly ban "near-miss" behavior.

Similarly, pseuCo programs allowing cycles are typically incorrect and will fail verification (because they permit an execution that spins without making progress, therefore never completing the step checklist). To speed up execution and provide better error messaging, unless otherwise configured, the verifier will identify cycles and abort verification with an error should one be found.

Syntactic Checking. In the Programming chapter of pseuCo Book, most exercises will want to control exactly how students use concurrency features. Part of this control is to ensure that students do not use shared memory in exercises about message passing, and vice versa.

This can, of course, be handled semantically by failing verification when an unauthorized concurrency feature is used. However, this is unnecessarily slow and complicated. A simpler approach, implemented in pseuCo Book, is to run an additional suite of syntactic checks when requested by the exercise, forbidding message passing or shared memory constructs as needed.

These checks run even before the user requests validation, after parsing and type-checking, allowing violations to be shown live in the editor UI, just like other types of syntactical errors. This also helps create the impression that the message passing and shared memory parts of pseuCo constitute different dialects of the language that cannot be freely mixed.

Banning message passing is done by simply banning any declaration of channel variables. Detecting and preventing shared memory is slightly more involved:

- Global declarations are banned, unless they are channels, and global channel variables cannot be assigned to.
- Locks and monitors cannot be declared, and join() statements are forbidden, all of which are considered shared-memory features.
- Procedures that take structures and arrays cannot be started. This prevents usage of shared memory by sharing pointers to heap-stored data structures across thread boundaries.
- Methods of structs cannot be started, and start() calls cannot be within a struct. This prevents sharing the implicit reference to the structure ("this-reference") between threads.

Together, these rules ensure that the only data that can be shared between agents directly is read-only global channel declarations.

Firing Set Validation and Data Race Detection. Most exercises in the shared memory section want to disallow *data races*. A data race occurs when a program can access a variable by two agents in parallel, with at least one access being a write access.

In the Petri net semantics of pseuCo, a data race is a marking that permits firings that are *conflicting*, i.e. (a) they represent actions taken by different agents, (b) they encode a global variable access to the same variable, (c) at least one access is writing, and (d) they access different *paths*.[3]

Formally, this can be made accessible from LTL by introducing a new atomic proposition that is applied to firings originating from markings that permit conflicting firings.

In the implementation, this is handled by extending the verification algorithm: In addition to the step checklist, a function *firingSetAllowed* can be passed to the verifier. For each marking of the Petri net, after evaluating the set of enabled firings, this function can inspect this set and may reject the combination of firings. To prevent data races, if enabled by the exercise, pseuCo Book applies a *firingSetAllowed* function that identifies conflicting firings and, if one is found, triggers a verification failure with an error message explaining the data race.

---

[3] This allows e.g. concurrent access to different indices of arrays, or different fields of a struct, despite these being stored within the same variable internally.

Restricted Actions. In some exercises, the exercise designer wants to strictly control the use of certain language features, e.g. message passing. To allow expressing this easily, without needing the full power of LTL, groups of atomic propositions can be declared as *restricted* in an exercise. Firings of the Petri net that use any restricted action are permitted only if they are required by the step checklist, i.e. advance the program's progress through the checklist.

Testing Mode. In practice, a small percentage of exercises in pseuCo Book permit solutions that exceed verification times that we consider acceptable in a teaching context (about 10 s on a reasonably modern computer and browser). For these exercises, at the expense of soundness, the framework can be configured to use *testing mode*. This replaces the exhaustive search with a fixed number of random walks, providing a compromise between a reasonably safe assurance a student's solution is correct and verification times.

4   Programming in pseuCo Book

Using the verification technology discussed in the previous section, we have expanded pseuCo Book to include a new chapter on Programming with pseuCo. Targeted at students that are already familiar with single-threaded programming in languages like Java or C, this chapter does not introduce any programming basics, but assumes it is a student's first contact with practical *concurrent* programming. It provides programming exercises that guide students through their first uses of all relevant concurrency features and concepts covered in the chapter, leaving more complex exercises to be covered in traditional exercise sheets of an accompanying lecture.

4.1   Structure

The chapter is structured into three main sections.

A Gentle Introduction to pseuCo. This section serves to introduce students to the pseuCo syntax and the basics of concurrency. It starts with a discussion of a simple, sequential pseuCo program, before asking students to write a *Hello World* program (using a procedure call). Next, another example introduces the `start()` statement, teaching students how to use the most basic form of concurrency (without any communication between the agents). To prevent students from accidentally using shared memory, global variables are banned for these examples.

Message passing is generally considered to be easier to use *correctly*, but to a naïve student, simply allowing shared global variables may appear easier. To motivate the need for more controlled communication mechanisms between threads, the introduction section closes with an "intermezzo" section explaining *The Dangers of Uncoordinated Access to Shared Memory*. First, this subsection asks students to write a variant of a *Hello World* program that is multi-threaded, using a shared string variable to "send" a greeting message to a worker thread that reads and prints it. To analyze whether such a program is correct, pseuCo Book then discusses a traditional *concurrent counting* example, using two agents to repeatedly decrement a variable using a postfix expression (n--). Students are encouraged to analyze this program in pseuCo.com and asked to determine the possible outputs. Because the postfix decrement operation is not atomic, this example can produce surprising results. PseuCo Book discusses this, and the general risks associated with data races, concluding this section.

Message Passing Concurrency. This section introduces channels, beginning with synchronous (handshaking) channels. After a brief explanation of their syntax and use, students are asked to write a first minimal example, using a string channel to assemble and print a greeting in a worker agent.

Then, pseuCo Book introduces asynchronous (buffered) channels. To demonstrate their ability to store messages, students are asked to write a similar program than before, but this time writing a message to a channel *before* starting the agent that retrieves it.

Then, this section finishes introducing the essential message passing features by explaining the use of the `select case` statement as discussed in Sect. 2.1.

To provide students some guidance on how programs using message passing concurrency can be structured, this section concludes by introducing two standard concepts:

Producer & Consumer: To introduce the producer-consumer-pattern, pseuCo Book focuses on an example program that computes the series of factorials of prime numbers, i.e. 2!, 3!, 5!, 7!, and so on. Students are given a sequential implementation and are then asked to parallelize this by splitting prime finding and factorization into two agents, with an asynchronous channel in between. This constitutes a producer-consumer pattern with a single producer and consumer each. The section discusses this, as well as the implications of adding more producers and consumers.

Pipelining: To explain pipelining, this subsection focuses on prime generation. Assuming all primes up to $\sqrt{n}$ are already known, primes up to n can be found by testing whether any of the smaller primes is a factor of the number in question. This lends itself well to a pipelining approach where each candidate number is passed from agent to agent, with each agent eliminating multiples of one specific prime. The students are asked to implement this, writing a program that uses prime sieving agents for 2, 3, and 5 to identify all primes between 6 and 25.

Shared Memory Concurrency. This section lifts the restrictions on global variables and sharing references to heap-stored data and gives students the tools and knowledge to control the problems this causes.

After a brief reminder of the dangers of uncoordinated shared memory already explained in the intermezzo section before the introduction of message passing, this section begins by introducing locks. Starting with a discussion of the features and correct usage of locks, the section revisits the concurrent counting example seen previously. In an exercise, students are asked to modify the program by adding locks to make it safe.

Next, pseuCo Book discusses reentrancy, i.e. the feature of locks allowing safely calling methods while already holding a lock needed by the callee, by allowing locks to be *re-entered* by the same thread as often as needed.

Then, pseuCo Book introduces arrays and structs. While these data structures, by themselves, are not related to concurrency, they introduce a potential source of hidden data races: In pseuCo, these data structures are stored on the heap. This means that a data race can be created without unsafely sharing a pseuCo variable directly, by copying a reference to an array or heap and then using both copies to access the same field on the heap concurrently. To illustrate this point, an exercise asks students to deliberately write a pseuCo program that has a data race without using any global variables.

Finally, pseuCo Book introduces monitors, i.e. data structures that handle protecting their data internally and can thus be used concurrently without additional protection from the outside.

In many languages, monitors are an implicit construct (e.g. achieved in Java by writing a class where all fields are **private** and all **public** methods are **synchronized**). PseuCo Book begins by introducing this concept abstractly, then asks students to apply it manually to a data structure called **MessageBox** implementing a simple, shared storage for a single integer. The corresponding exercise asks students to manually add a lock to a given template implementation of **MessageBox**, then write a sample program that uses two agents where one writes a message to the box and the other uses polling to retrieve the message as soon as possible.

In pseuCo, a **monitor** is an explicit language feature, automating the process of protecting all methods of a structure. A pseuCo **monitor** is similar to a **struct**, with the following differences:

- When a monitor is instantiated, an implicit, managed lock is initialized.
- The lock is acquired and returned automatically on all entry and exit paths of every method of the monitor.
- Monitors allow declaring and using **conditions**, allowing condition synchronization, i.e. waiting and signalling similar to e.g. Java's **wait()** and **notify()** mechanism.

PseuCo **struct**s already do not permit direct access to fields, so no change is needed in this regard.

After a brief explanation of pseuCo's **monitor** features, pseuCo Book discusses the problems associated with busy waiting (as used by the students in the previous exercise) and attempting to wait for another agent to make a change to a data structure while holding the lock to it. To finish this section, students are then asked to implement a **monitor**-based **MessageBox** (by adding condition synchronization to an otherwise complete template).

4.2 Exercises and Verification

Overall, the *Programming* chapter of pseuCo Book contains 11 programming exercises, spread throughout the chapter and integrated into the narrative.

10 of these exercise use true verification, with verification runtimes not exceeding 10 s on reasonably modern systems for the intended solutions. Some incorrect solutions can create longer runtimes or cause the verifier to diverge, e.g. when containing an infinite loop that changes the program state. If verification takes longer than a pre-determined warning threshold, students are informed that the process "is taking longer than usual" and are asked to check for infinite loops and try to simplify their program. This is merely a warning – students can wait for verification to finish, or cancel at any time. (As verification is run on the student's machine, there is no need to enforce a timeout).

A single exercise, the prime sieve, is set to testing mode (see Sect. 3.2) to combat high verification runtimes. It uses 100 random walks to provide reasonable assurance that a solution is correct. This is also shown in the UI.

Most exercises disallow infinite loops as a violation of the specification. A single exercise, the student's first attempt to manually create a monitor-like structure called `MessageBox`, allows – and in fact requires – the presence of an infinite loop. This does not pose any technical difficulty for verification as this loop does not significantly increase the size of the state space.

All programming exercises share the same user interface and logic. Therefore, adding a new exercise to pseuCo Book does not require custom code, only

- an exercise configuration file for the frontend, describing
  - which pseuCo dialects (message passing/shared memory) are allowed,
  - the description of the exercise as shown to students, and
  - the template given to students when they begin (if any); and
- a verification background worker, i.e. a verification configuration containing
  - the property to analyze, i.e. the step checklist, composing pre-made groups of atomic propositions,
  - any additional checks (e.g. syntax checks or firing set validators), and
  - whether to use testing mode and if so, the number of traces to check.

This is demonstrated in Listings 1.2 and 1.3, showing the internal definition of the "Hello Message Passing World!" exercise.

Listing 1.2. Exercise configuration for the "Hello Message Passing World!" exercise

```
1  const config: PseuCoProgrammingConfiguration = {
2      allowedDialects: { // configure text editor
3          mp: true,      // message passing = OK
4          sm: false      // shared memory = syntax error
5      },
6      exerciseDescription: <div>
7          <p>Write a pseuCo-MP program that prints <code>"Hello,␣World!"</code>
               ↳ by sending <code>"World"</code> on a synchronous channel to an
               ↳ agent that assembles and prints the greeting.</p>
8      </div>,
9      getWorker: () => new Worker(new URL('./worker.ts', import.meta.url))
10                      // call this worker for verification
11 };
```

Listing 1.3. Definition of the background worker handling verification in the "Hello Message Passing World!" exercise (see also Eq. (4))

```
1  const validatorConfiguration: FlowReachabilityGraphValidatorConfiguration = {
2      steps: {
3          order: "sequential",
4          steps: [{
5              moveCompletesStep: moveValidatorStartAgent(),
6              description: 'start an agent'
7          }, {
8              moveCompletesStep: moveValidatorAnd(
                 ↳ moveValidatorParticipatingAgents([0, 1]),
                 ↳ moveValidatorHandshaking((v) => /^World!?$/i.test(v.toString()
                 ↳), false)),
9              description: 'send "World" from the main agent to the first
                 ↳ started agent on a synchronous channel'
10         }, {
11             moveCompletesStep: moveValidatorAnd(
                 ↳ moveValidatorParticipatingAgents([1]), moveValidatorPrintLn(/^
                 ↳ Hello,? World!?$/i, false)),
12             description: 'print "Hello, World!" (from the first agent)'
13         }]
14     },
15     restrictedMoves: [{ // no message passing except as required above
16         detector: moveValidatorMessagePassing(),
17         description: "performed a message-passing action"
18     }]
19 };
20 registerValidationWorkerCallback(flowReachabilityGraphValidator(
     ↳ validatorConfiguration), { mp: true, sm: false });
```

5 Conclusion

In this paper, we have developed an extension of pseuCo Book: A chapter on concurrent programming, designed as the first contact point of students with concurrency in practical programming. It uses pseuCo, a programming language specifically designed for teaching, to introduce both message passing and shared memory concurrency features. Integrated programming exercises help give students their first experiences with concurrency in a controlled environment. An integrated autograder, running directly inside the web app on students' devices, implements LTL-based model checking, backed by pseuCo's Petri net semantics. It relies on a set of rich, compiler-generated labels on the Petri net. This enables more than verifying the externally visible behavior of the program: It grants a deep insight into the internal workings of the program, ensuring the programming exercises are solved using the concurrency control mechanisms specified by the exercise designer. The new chapter of pseuCo Book is freely accessible at https://book.pseuco.com/#/read/pseuco/.

Acknowledgements. This work was partially funded by the Deutsche Forschungsgemeinschaft (DFG, German Research Foundation) – project number 389792660 – TRR 248 – CPEC, see https://perspicuous-computing.science. The author would like to thank Holger Hermanns for his contributions.

References

1. Biewer, S., Freiberger, F., Held, P.L., Hermanns, H.: Teaching academic concurrency to amazing students. In: Aceto, L., Bacci, G., Bacci, G., Ingólfsdóttir, A., Legay, A., Mardare, R. (eds.) Models, Algorithms, Logics and Tools. LNCS, vol. 10460, pp. 170–195. Springer, Cham (2017). https://doi.org/10.1007/978-3-319-63121-9_9
2. Bruzual, D., Freire, M.L.M., Di Francesco, M.: Automated assessment of Android exercises with cloud-native technologies. In: Proceedings of the 2020 ACM Conference on Innovation and Technology in Computer Science Education, ITiCSE 2020, Trondheim, Norway, 15–19 June 2020, pp. 40–46. ACM (2020). https://doi.org/10.1145/3341525.3387430
3. Carbunescu, R., et al.: Architecting an autograder for parallel code. In: Annual Conference of the Extreme Science and Engineering Discovery Environment, XSEDE 2014, Atlanta, GA, USA, 13–18 July 2014, pp. 68:1–68:8. ACM (2014). https://doi.org/10.1145/2616498.2616571
4. Freiberger, F.: pseuCo Book: an interactive learning experience. In: ITiCSE 2022: Innovation and Technology in Computer Science Education, Dublin, Ireland, 8–13 July 2022, vol. 1, pp. 414–420. ACM (2022). https://doi.org/10.1145/3502718.3524801
5. Freiberger, F., Hermanns, H.: Concurrent programming from PSEUCO to Petri. In: Donatelli, S., Haar, S. (eds.) PETRI NETS 2019. LNCS, vol. 11522, pp. 279–297. Springer, Cham (2019). https://doi.org/10.1007/978-3-030-21571-2_16
6. Gastin, P., Oddoux, D.: Fast LTL to Büchi automata translation. In: Berry, G., Comon, H., Finkel, A. (eds.) CAV 2001. LNCS, vol. 2102, pp. 53–65. Springer, Heidelberg (2001). https://doi.org/10.1007/3-540-44585-4_6
7. Gerth, R., et al.: Simple on-the-fly automatic verification of linear temporal logic. In: Protocol Specification, Testing and Verification XV, Proceedings of the Fifteenth IFIP WG6.1 International Symposium on Protocol Specification, Testing and Verification, Warsaw, Poland, June 1995. IFIP Conference Proceedings, pp. 3–18. Chapman & Hall (1995)
8. Jensen, K.: Coloured Petri Nets - Basic Concepts, Analysis Methods and Practical Use, vol. 1. Springer, Heidelberg (1992)
9. Pnueli, A.: The temporal logic of programs. In: 18th Annual Symposium on Foundations of Computer Science, Providence, Rhode Island, USA, 31 October–1 November 1977, pp. 46–57. IEEE Computer Society (1977). https://doi.org/10.1109/SFCS.1977.32
10. Stahlbauer, A., Kreis, M., Fraser, G.: Testing scratch programs automatically. In: Proceedings of the ACM Joint Meeting on European Software Engineering Conference and Symposium on the Foundations of Software Engineering, ESEC/SIGSOFT FSE 2019, Tallinn, Estonia, 26–30 August 2019, pp. 165–175. ACM (2019). https://doi.org/10.1145/3338906.3338910
11. Wang, W., et al.: SnapCheck: automated testing for Snap! Programs. In: ITiCSE 2021: Proceedings of the 26th ACM Conference on Innovation and Technology in Computer Science Education vol 1, Virtual Event, Germany, 26 June –1 July 2021, pp. 227–233. ACM (2021). https://doi.org/10.1145/3430665.3456367

Teaching TLA$^+$ to Engineers at Microsoft

Markus A. Kuppe$^{(\boxtimes)}$ ⓘ

Microsoft Research, Redmond, USA
makuppe@microsoft.com

Abstract. This work presents an effort to promote and teach modeling and formal methods to software engineers at Microsoft. The class focuses on a practical application of the TLA$^+$ specification language, using it to model a termination detection algorithm and verify correctness properties with an explicit-state model checker. The auxiliary learning material introduces symbolic model checking and theorem proving. The class emphasizes the application of modeling with TLA$^+$ rather than a formal introduction of concepts. As a result of completing the class, software engineers will gain proficiency in using TLA$^+$ and its relevant language constructs. The results of a survey suggest that the class successfully promotes the use of modeling and formal methods at Microsoft.

1 Introduction

Formal methods experts and tool builders would not dare to build large systems without having the most rigorous approaches at their disposal. Yet, in practice, modeling and formal verification are largely ignored by software engineering practitioners[1]. How can we, the formal methods experts, convince our peers in the industry to adopt our formalisms and tools to build more robust and thus cheaper-to-operate systems?

One part of the answer is demonstrating the usefulness of modeling and formal methods. Another part is educating software engineers on our formalisms and tools. To address these two challenges, the author created and regularly teaches a class that targets software engineers with little-to-no prior exposure to modeling and formal methods.

The modeling language taught is the TLA$^+$ specification language. TLA$^+$ is a high-level, math-based, formal specification language that is used to design, specify, and document systems. A specification describes a state machine and is specified by formulas expressed in the Temporal Logic of Actions [7,8]. TLA$^+$ is an untyped language in which data structures are represented by Zermelo-Fraenkel set theory with choice. TLA$^+$ is implementation language agnostic and is used to find bugs above the code level. Users can check and reason about TLA$^+$ specs with the explicit-state model checker TLC, the symbolic model checker Apalache, and the TLA$^+$ proof system (TLAPS). While TLC and Apalache are used to check finite models, TLAPS supports deductive reasoning [1,3,10].

---

[1] Unless forced by external requirements in safety-critical domains.

C. Dubois and P. San Pietro (Eds.): FMTea 2023, LNCS 13962, pp. 66–81, 2023.
https://doi.org/10.1007/978-3-031-27534-0_5

Convincing software engineers of the usefulness of modeling and formal methods is an important objective of this class. Given that designing a novel algorithm from scratch is ill-suited for an introductory class due to its many unknowns, the author instead decided to study a known distributed termination detection algorithm [2]. Students specify the *EWD*998 algorithm in TLA+, define suitable fairness constraints, verify safety and liveness properties, and encounter refinement during the two-day, hands-on class.

2   How We Teach TLA+

In this section, we outline the setup and rationale and describe a typical progression of the two-day class enriched with the author's observation. A more objective evaluation of the class is reserved for Sect. 3.

2.1   Prerequisites and Preliminaries

Before the class, students are asked to watch Lamport's "*Introduction to TLA+*" video [9], which provides an overview of the concept of a state machine, and to read Dijkstra's paper on *EWD*998. However, we have found that not all students have the time to complete these preparatory tasks.

The first day of the class begins with a brief introduction to TLA+ and an overview of the class structure and expectations. We remind students that this class is not graded and encourage them to actively participate in discussions. We also let them know that the instructor will intentionally make mistakes during group exercises as a teaching method to keep students engaged and facilitate learning through discussion.

We provide students with access to a full TLA+ IDE and the three verification tools (TLC, Apalache, and TLAPS) via the VSCode extension, which can be easily installed in cloud-based development environments like GitHub Codespaces or GitPod[2]. This ease of installation is important given the class' time constraints and the reluctance of some engineers to prepare in advance. Ease of installation is also why we chose VSCode over the more mature TLA+ Toolbox [6].

The class learning materials are provided in the form of a source code repository rather than prose to build on the familiarity of engineers with this format and to mimic real-world modeling. Each commit in the repository corresponds to a learning step and includes extensive comments that capture the verbal explanations of the instructor. This allows students to review the material offline and also ensures that they can start from a known good state when working independently after group discussions. The learning material is available under the permissive MIT license at https://github.com/tlaplus-workshops/ewd998.

---

[2] https://github.com/features/codespaces and https://www.gitpod.io.

2.2 Problem Description

As hinted in Sect. 1, the decision to use *EWD*998 was based on several criteria, including the fact that it is a distributed system, has moderate complexity because of reliable message delivery and no node failures, and its age makes it relatively unknown to the students participating in the class.

In order to simulate the process of spec writing with TLA$^+$ in a real-world scenario, students are given an informal description of the *EWD*998 algorithm and an animation of its execution[3]. The description outlines a distributed system in which nodes are organized in a ring and one node is designated as the leader. The leader sends a token around the ring, and when it is returned, it is assumed that all other nodes have completed their computations. However, the description does not account for asynchronous message delivery, in which a node can be reactivated by a message from another node. This omission is pointed out by the instructor, and the description is amended to include a message counter and tally of in-flight messages recorded on the token. Occasionally, the instructor also lets the students enact an execution of the algorithm. The purpose of this exercise is to encourage students to critically think about the algorithm and identify any omissions, leading into the use of TLA$^+$ for modeling and formal verification.

2.3 Towards a High-Level State Machine

In order to introduce students to TLA$^+$ and its syntax, the class begins with a review of set theory. Using the REPL, the students practice defining sets and performing common operations, such as intersections, unions, and set minus. They then apply this knowledge by declaring the first spec constant and defining a corresponding assumption. It is noted that students do not typically challenge or question the fact that TLA$^+$ is typeless when learning about it.

In the next step of the process, the students study an animation of the *EWD*998 algorithm to identify state changes. The animation provides more information than what engineers typically know in the early stages of software development, but this compromise is made in the interest of time. The students typically identify the nodes, in-flight messages, the token, and the node and token state. However, only the variables *active* and *pending* are codified in the first spec (see Sect. 5). The students are told that the concept of refinement allows them to model the token, the node counters, and the color at a later stage. The two variables, *active* and *pending*, are modeled as TLA$^+$ functions but are called arrays for familiarity. The fact that TLA$^+$ has only global variables is also discussed, as this can lead to the accidental access of a remote state (compare Listing 6 for an example), challenging the students' understanding of distributed systems as decompositions into a node-local state.

---

[3] https://github.com/tlaplus-workshops/ewd998/blob/main/figures/v01-ring04.gif.

Listing 1 Evolution of the *Terminate* action from pseudocode to TLA$^+$.

(Incorrect Formula) Pseudocode with an implicit program counter and assignment.
$Terminate(node) \triangleq active[node] = $ FALSE

(Incorrect Formula) An action is true or false of a pair of states, i.e., it is a transition in a state machine. Introduction of the prime operator.
$Terminate(node) \triangleq active[node] = $ TRUE $\wedge$ $active[node]' = $ FALSE

(Incorrect Formula) Junction list syntax: a list of sub-formulas prefixed by / or and indentation is used to eliminate paretheses.
$Terminate(node) \triangleq$
$\wedge active[node] = $ TRUE
$\wedge active[node]' = $ FALSE

(Incomplete Formula) TLA+ is typeless; defines value of active' to be a function.
$Terminate(node) \triangleq$
$\wedge active[node] = $ TRUE
$\wedge active' = [n \in Node \mapsto$ IF $n = node$ THEN FALSE ELSE $active[n]]$

(Proper Formula) TLA requires an action to define the value of all variables.
$Terminate(node) \triangleq$
$\wedge active[node] = $ TRUE
$\wedge active' = [n \in Node \mapsto$ IF $n = node$ THEN FALSE ELSE $active[n]]$
$\wedge pending' = pending$

(Proper Formula) Syntax sugar and removal of superfluous enablement condition, i.e., a Terminate action of an inactive node leaves all variables unchanged.
$Terminate(node) \triangleq$
$\wedge active' = [active$ EXCEPT $![n] = $ FALSE$]$
$\wedge$ UNCHANGED $pending$

The next task for the students is to describe the actions of the *EWD998* algorithm verbally. This verbal description is first turned into imperative pseudocode with assignments and then gradually converted into TLA$^+$, with the process of conversion shown in Listing 1. It is most effective to teach students that in TLA$^+$ an action is a mathematical formula rather than a piece of code with separately evaluated sub-expressions. This helps shift the focus to state machines, in which actions (transitions) are atomic and do not interleave. We also cover preconditions, also known as action *enablement* in TLA$^+$ terminology, and understand that removing the enablement condition for the *Terminate* action does not result in new states in the state graph.

After working in a group to define the *Terminate* action, the students individually define the *SendMsg* and *Wakeup* actions (lines 16 to 21 in Listing 5). Defining the initial state of the state machine, represented by the initial predicate *Init*, is straightforward once TLA$^+$ actions have been discussed. At this point, the students notice that all three actions declare parameters that are not

Listing 2 Evolution of the next-state relation *Next* from an incorrect conjunct list to modeling non-determinism with disjunction, to existential quantification ranging over the spec's *Nodes* constant (syntactically compressed to preserve space).

(Wrong Formula) Attempt to model that Terminate, Wakeup, and SendMsg actions may happen. Next is a constradiction causing TLC to report one distinct state.
$Next \triangleq$
$\quad \wedge\ Terminate(0)\ \wedge\ Terminate(1)\ \wedge\ Wakeup(0) \wedge\ Wakeup(1)$
$\quad \wedge\ SendMsg(1, 0) \wedge SendMsg(0, 1) \wedge SendMsg(0, 0) \wedge SendMsg(1, 1)$

(Proper Formula) Model Terminate, Wakeup, or SendMsg actions to happen non-deterministically for any of the two nodes.
$Next \triangleq$
$\quad \vee\ Terminate(0)\ \vee\ Terminate(1)\ \vee\ Wakeup(0) \vee\ Wakeup(1)$
$\quad \vee\ SendMsg(1, 0) \vee SendMsg(0, 1) \vee SendMsg(0, 0) \vee SendMsg(1, 1)$

(Proper Formula) Model Terminate, Wakeup, or SendMsg actions to happen non-deterministically for any element of the set of nodes Nodes.
$Next \triangleq$
$\quad \exists\, n,\ m \in Node :$
$\qquad Terminate(n) \vee Wakeup(n) \vee SendMsg(n, m)$

defined. This is when the next-state relation, *Next*, is defined (see Listing 2). Instead of immediately using predicate logic, *Next* is initially defined as a conjunct list for two nodes. This results in TLC finding no counterexample and generating only a single (initial) state.

Changing conjunction to disjunction allows students to study non-determinism without introducing existential quantification. At this point, we also briefly explain the state-space explosion problem and the tradeoffs of Apalache and TLAPS. Afterward, we use a TLC state constraint involving universal quantification to bind the state space (line 28 in Listing 5).

Note that we introduce new language gradually to avoid overwhelming the students with too much new syntax. However, a language cheat sheet is provided for reference[4].

2.4 Invariants

The instructor introduces a bug in the Wakeup action to focus on verification, leading to a discussion of arrays, functions, and sets of functions and introducing the TypeOK invariant (compare Listing 3). When students check the invariant, they see their first counterexample, which demonstrates the value of TLA[+] in providing complete traces of variables without the need for explicit log statements. This is a compelling argument in favor of modeling for engineers who typically debug systems without complete logs.

---

[4] https://lamport.azurewebsites.net/tla/summary-standalone.pdf.

Listing 3 Excerpt with the incorrect *Wakeup* action and the *TypeOK* invariant.

$TypeOK \triangleq$
    $active \in [Node \rightarrow \text{BOOLEAN}] \land pending \in [Node \rightarrow Nat]$

(Wrong Formula)
$Wakeup(i) \triangleq \land pending[i] > 0$
    $\land active' = [active \text{ EXCEPT } ![i] = \text{TRUE}]$
    $pending' = [pending \text{ EXCEPT } ![i] = @ - 1]$
    $\land pending' = [pending \text{ EXCEPT } ![i] = @ - 2]$ Bug A: @-2

Listing 4 Excerpt showing the amended *Terminate* action and *Stable* invariant.

The system terminated when all nodes are inactive and no messages are in flight.
$terminated \triangleq \forall n \in Node : \neg active[n] \land pending[n] = 0$

. . .

$Terminate(i) \triangleq \land active' = [active \text{ EXCEPT } ![i] = \text{FALSE}]$
    $\land \text{UNCHANGED } pending$
    $\land \lor \text{UNCHANGED } terminationDetected$
        $\lor terminationDetected' = \text{TRUE}$ Bug B discussed in Section 2.5.
        $\lor terminationDetected' = terminated'$ Bug C discussed in Section 2.5.

$Stable \triangleq terminationDetected \implies terminated$

To model high-level termination detection, the instructor introduces the new *terminationDetected* variable and demonstrates the legality of priming any state-level formula in TLA+. A new invariant *Stable* is initially defined as *Stable* $\triangleq$ *IF terminationDetected THEN terminated ELSE TRUE*, but it is later refined to *Stable* $\triangleq$ *terminationDetected* $\Rightarrow$ *terminated* to introduce implication separately (Listing 4).

To reinforce the previous lessons, the class works on a TLA+ specification for a logic puzzle inspired by the movie Die Hard.[5] The problem description is straightforward, and students are able to complete the specification independently, with the instructor available to answer questions. This exercise is especially helpful for weaker students.

2.5 Safety and Liveness

We could end the class after the previous section and leave temporal logic, including safety and liveness properties, fairness, and refinement, for a 201 class on TLA+. However, we find it essential to discuss the more advanced concepts of TLA+ to make software engineers aware of their existence.

To illustrate safety properties, we modify the definition of the *Terminate* action by adding the conjunct labeled *Bug B* (see Listing 4) and check the *Stable*

---

[5] https://github.com/tlaplus/Examples/blob/master/specifications/DieHard/
DieHard.tla.

Listing 5 A (lasso-shaped) counterexample violating the property *Live* $\triangleq$ $\diamond terminationDetected$.

1: <Initial predicate>
$\wedge\ pending = (0 :> 0 @@ 1 :> 0)$
$\wedge\ active = (0 :>$ FALSE $@@ 1 :>$ TRUE$)$
$\wedge\ terminationDetected =$ FALSE

2: <SendMsg of module AsyncTerminationDetection>
$\wedge\ pending = (0 :> 1 @@ 1 :> 0)$
$\wedge\ active = (0 :>$ FALSE $@@ 1 :>$ TRUE$)$
$\wedge\ terminationDetected =$ FALSE

3: <Wakeup of module AsyncTerminationDetection>
$\wedge\ pending = (0 :> 0 @@ 1 :> 0)$
$\wedge\ active = (0 :>$ TRUE $@@ 1 :>$ TRUE$)$
$\wedge\ terminationDetected =$ FALSE

1: Back to state: <Terminate of module AsyncTerminationDetection>

formula as a safety property rather than an invariant. This often surprises students, as the current version of the *Stable* property is only true for initial states, and it must be strengthened to *Stable* $\triangleq \Box\ (terminationDetected \Rightarrow terminated)$ for TLC to find a violation.

To demonstrate the concept of liveness, we have students reflect on the guarantees of the property *Stable*. To facilitate this discussion, the instructor intentionally breaks the specification by removing the disjunct that ensures the eventual termination detection (see *Bug C* in Listing 4). This change does not violate the safety property *Stable* because *Stable* does not define the "good thing that must eventually happen."

Formally introducing the concept of liveness is challenging due to its connection with fairness. Checking a liveness property such as *Live* $\triangleq$ $\diamond terminationDetected$ without defining fairness would cause TLC to find a somewhat unexpected and distracting counterexample of infinite stuttering after the initial state. Thus, we add the behavior specification *Spec* $\triangleq Init \wedge \Box\ [Next]_{vars} \wedge$ $WF_{vars}\ (Next)$ but delay its proper introduction until later.

Checking *Spec* $\Rightarrow$ *Live* reveals a lasso-shaped counterexample where termination is never detected because the system never terminates in the first place, which makes us realize that *Live* is too strong and has to be weakened to *Live* $\triangleq \Box\ (terminated \Rightarrow \diamond terminationDetected)$. Finally, we verify that the invariant *TypeOK* and the two properties *Stable* and *Live* hold for the *AsyncTerminationDetection* (*ATD*) specification for three and four nodes. If students question the verification guarantees of model checking, we refer them to a dedicated paper on formally verifying EWD998 with TLC, Apalache, and TLAPS [4].

2.6   Refinement with EWD998

In Sect. 2.3, we promised to model a more detailed specification of termination detection. Now that we have completed *ATD*, our students are ready to begin specifying *EWD*998. We instruct them to copy the *ATD* model and remove all occurrences of *terminationDetected*, replacing them with the rules below, which have been extracted from Dijkstra's description of the algorithm:

1. The initiator sends the token with a counter q initialized to 0 and color white.
2. The initiator starts a new round iff the current round is inconclusive.
3. The initiator whitens itself and the token when initiating a new round.
4. Sending a message by node n increments a counter at n, and the receipt of a message decrements n's counter.
5. Receiving a message (not token) blackens the (receiver) node.
6. An active node m—owning the token—keeps the token. If m becomes inactive, it passes the token to its neighbor with $q = q + counter[m]$.
7. A black node taints the token.
8. Passing the token whitens the sender node.

This exercise is a key part of the course and helps students solidify their understanding of both TLA$^+$ and the *EWD*998 algorithm. As a result of this exercise, students will add additional variables, make modifications to the *SendMsg* and *Wakeup* actions, and create two new actions called *InitiateToken* and *PassToken*. During the exercise, students will also learn how to model the *token* as a record, a standard TLA$^+$ function with a domain of strings (compare lines 31 to 65 in Listing 5).

Students may notice that the definitions of *Stable* and *Live* break after the removal of *terminationDetected*. This serves as an opportunity to introduce the concept of refinement by stating that the set of behaviors defined by the *EWD*998 spec is a subset of the behaviors defined by *ATD*. We then demonstrate how to define and check refinement with *terminationDetected* substituted for *TRUE* or *FALSE* (see the first two variants in Listing 6), for which TLC reports a safety and a liveness violation, respectively. It is not uncommon for students to include global state in the refinement mapping, as in the third version of *ATD* in Listing 6. Students discover the correct refinement mapping with the help of TLC.

TLC may report subtle liveness violations if *ATD!Spec* has a fairness conjunct such as $WF_{vars}(Next)$. In class, we help students by pointing them towards the correct fairness constraints for *ATD* and *EWD*998. However, we leave it as an exercise for the reader to find the correct fairness constraint in this case.

2.7   Fairness

In past installments of the class, we attempted to cover both weak and strong fairness by deriving them from their constituent formulas. This involved introducing:

Listing 6 Evolution of the refinement mapping in spec $EWD998$.

(Wrong Refinement Mapping) Safety violation

$ATD \triangleq$ INSTANCE $AsyncTerminationDetection$ WITH $terminationDetected \leftarrow$ TRUE

(Wrong Refinement Mapping): Liveness violation

$ATD \triangleq$ INSTANCE $AsyncTerminationDetection$ WITH $terminationDetected \leftarrow$ FALSE

(Wrong Refinement Mapping) Mapping includes global system state.

$ATD \triangleq$ INSTANCE $AsyncTerminationDetection$ WITH $terminationDetected \leftarrow$
$\quad \forall n \in Node : \neg active[n] \land pending[n] = 0$

(Proper Refinement Mapping): Only state of node zero (initiator) is included.

$ATD \triangleq$ INSTANCE $AsyncTerminationDetection$ WITH $terminationDetected \leftarrow$
$\quad \land \quad \neg active[0] \land token.pos = 0 \land token.color = $ "white"
$\quad \land \quad token.q + counter[0] = 0 \land color[0] = $ "white"

Listing 7 Choosing a receiver of a message in $SendMsg$ with the $CHOOSE$ operator.

$SendMsg(i) \triangleq$
$\quad \land \quad active[i] \land$ UNCHANGED $\langle active, color, token \rangle$
$\quad \land \quad counter' = [counter$ EXCEPT $![i] = @ + 1]$
$\quad$ E recv $\in$ Node: pending' $= [pending$ EXCEPT $![recv] = @ + 1]$
$\quad \land pending' = [pending$ EXCEPT $![$CHOOSE $n \in Node : n \neq i] = @ + 1]$

- $ENABLED\ A$ (with A an action-level formula)
- $[A]_v \iff A \lor v = v'$
- $\langle A \rangle_v \iff A \land v \neq v'$
- $\Box \diamond A$ ("eventually always")
- $\diamond \Box A$ ("always eventually")

However, we found that these concepts are beyond the scope of the class due to time constraints. Today, we refer students to Lamport's book *Specifying Systems* or the corresponding lecture in his video course [8] for an in-depth discussion on these concepts. It is worth noting that the AB protocol in Lamport's video lecture requires strong fairness in order for its liveness property to hold, whereas $EWD998$ does not.

2.8 Deterministic Choice (CHOOSE)

The $CHOOSE$ operator is commonly used in everyday TLA$^+$ modeling. It is known to mathematicians as Hilbert's ϵ-operator. While we do not discuss its formal properties with students, we do want to mention a common pitfall related to $CHOOSE$. If the message receiver $recv$ in the action $SendMsg$ is chosen deterministically rather than non-deterministically (as shown in Listing 7), the model's state space is significantly reduced. Students can observe this reduction by studying TLC's state-space statistics. At this point, we also

Listing 8 Intentionally added bug: unconditionally transfer the node's *color* to the *token*.

$PassToken(i) \triangleq$
  $\land \neg active[i] \land token.pos = i \land$ UNCHANGED $\langle active, pending, counter \rangle$
  $\land token' = [token$ EXCEPT $!.pos = @ - 1, \quad !.q \quad = @ + counter[i],$
              $!.color =$ IF $color[i] =$ "black" THEN "black" ELSE @]
              $!.color = color[i]]$ Unconditionally color token.
  $\land color' = [color$ EXCEPT $![i] =$ "white"]

---

demonstrate the TLA$^+$ debugger, which supports interactive state-space exploration [5]. It is important to note that this use of *CHOOSE* can cause the verification to miss bugs, so we strongly alert students to this issue. To further explain the semantics of *CHOOSE*, we evaluate different example expressions with TLC's REPL. It can also be helpful to contrast *CHOOSE* with *TLC!RandomElement*, an operator that uniformly at random picks an element from a given set. Lastly, we ask students to come up with the definition of $Max(S) \triangleq CHOOSE \, s \in S : \forall t \in S : s \geq t$ to see when *CHOOSE* is required.

2.9   Inductive Invariant and Recursion

The final portion of the class focuses on the importance of being vigilant, as Lamport advises us to "always be suspicious of success." As part of this discussion, we remind the students that we have previously verified that the specification *EWD*998 satisfies the safety and liveness requirements of the *ATD* spec for systems with up to three nodes. However, if the definition of the *PassToken* action is flawed and the node's *color* is transferred onto the *token* unconditionally (see Listing 8), this bug may not be detectable with fewer than four nodes. When using model checking to verify refinement with four nodes, we are able to detect this bug, but TLC takes several minutes to find its ten-step counterexample, which may be too long for impatient students. Fortunately, Dijkstra's paper offers an inductive invariant that can be checked more efficiently with TLC's reachability checker. Thus, the students are tasked to translate Dijkstra's informal definition into TLA$^+$ and verify it with TLC:

P0: $(Si : 0 \leq i < N : p.i) = (Si : 0 \leq i < N : c.i)$
P1: $(Ai : t < i < N : machine \; nr.i \; is \; passive) \land (Si : t < i < N : c.i) = q$
P2: $(Si : 0 \leq i \leq t : c.i) + q > 0$
P3: $Ei : 0 \leq i \leq t : machine \; nr.i \; is \; black$
P4: The *token* is black
Inv: $P0 \land (P1 \lor P2 \lor P3 \lor P4)$

Translating *Inv* into TLA$^+$ is straightforward due to the similarities in syntax (Listing 9), but the students may need to learn about recursive functions, recursive operators, and folds in order to define Dijkstra's sum operator (S). When checked with TLC for four nodes, TLC quickly find the safety violation.

After completing this exercise, we open the floor for discussion, which often centers around applying TLA$^+$ to internal projects.

Listing 9 Dijkstra's inductive invariant in TLA$^+$, involving a recursive operator definition.

```
RECURSIVE Sum(_, _, _)
Sum(fun, from, to) ≜
        IF from = to THEN fun[to] ELSE fun[from] + Sum(fun, from + 1, to)

Inv ≜
        ∧ P0:: Sum(pending, 0, N − 1) = Sum(counter, 0, N − 1)
        ∧   ∨ P1:: ∧ ∀ i ∈ (token.pos + 1) .. N − 1 : ¬active[i]
                ∧ IF token.pos   = N − 1 THEN token.q = 0
                     ELSE   token.q = Sum(counter, (token.pos + 1), N − 1)
              ∨ P2:: Sum(counter, 0, token.pos) + token.q > 0
              ∨ P3:: ∃ i ∈ 0 .. token.pos : color[i] = "black"
              ∨ P4:: token.color = "black"
```

3   Demographics and Evaluation

Our two-day TLA$^+$ class is advertised through a Microsoft-internal learning platform in the "Azure Advanced" category, competing with other courses, such as system-level programming. The class is open to all Microsoft engineers, regardless of their seniority. On average, we teach one class per month, with the size of the class depending on the availability of a teaching assistant. In-person classes with one instructor are limited to 50 students, and virtual classes are generally restricted to 20 students. When a teaching assistant is present, the class size can be increased to 100 students. Attending the class is free of charge. So far, we have delivered more than 25 classes with a total of over 300 students. Spinoff versions of the class have also been held publicly at conferences such as Hydraconf 2021, Strange Loop 2021, and MCH 2022 for audiences of 20 to 70 people.

For Microsoft-internal classes, we have access to the following aggregated data about student demographics:

- The majority of students are software engineers, with hardware engineers, solution architects, and program managers being outliers.
- Around half of the registrants attend a class.
- The majority of students have recently joined Microsoft, and thus we believe that most students are junior engineers.

At the start of a class, students self-introduce by answering questions about their background and motivation for learning TLA$^+$. While we do not systematically record their answers, our observations are as follows:

- The majority of students attend the class out of personal interest.
- A small minority of students have previous experience with formal methods, and they tend to perform well in the class.
- There appears to be a positive correlation between seniority and an appreciation for modeling and verifying systems beyond the code level.

Lastly, we survey the students sometime after the class. The survey is intentionally kept short in order to increase the number of responses. However, we recognize that the survey may suffer from selection bias. The answers of 94 participants out of more than 200 students are presented below.[6] When the numbers do not add up to 94, it means that multiple answers are possible:

1. *Should the workshop provide warmup/preparation material such as a basic set theory refresher?*
   - *Yes (57), No (27), I studied Specifying Systems, Lamport's Video course, Practical TLA+, ... in advance (23)*
2. *Does the algorithm EWD998 strike the right level of complexity?*
   - *Prefer a more sophisticated problem such as Paxos (5), Keep EWD998 (70), A less sophisticated problem such as EWD840, ... (19)*
3. *Compress the workshop to a single day or extend the workshop to three days?*
   - *Compress to one day (7), Keep at two days (51), Extend to three or more days (36)*
4. *I would have preferred more or fewer hands-on exercises?*
   - *More group exercises (40), More non-group (individual) exercises (47), Fewer group exercises (7)*
5. *What I've learned about TLA+ has already been useful for my job?*
   - *Yes (47), No (47)*
6. *If you didn't finish the workshop, you dropped out because...*
   - *TLA+ doesn't seem relevant to me and my work (2), Had to tend to other work, such as a live-site incident (11), Got bored, workshop was too slow (1), Workshop was too fast for me (10)*

The aggregated survey responses suggest that providing optional warmup material and extending the class by an additional third day would be helpful. This would allow students to have more time to work on exercises and also allow them to begin writing work-related specifications with guidance from the instructors. In response to question 5, there is no correlation between students reporting that TLA+ is applicable to their job and their seniority (which was binned into junior, senior, and principal categories). However, there is a direct correlation between seniority and students finding TLA+ not useful; the more junior a student, the less value they see in TLA+. This aligns with the instructor's observation mentioned earlier.

---

[6] We did not send out the survey for the first few classes.

4   Related Work

The TLA$^+$ community maintains a list of university lectures that use TLA$^+$ or PlusCal to teach distributed and concurrent systems or introduce TLA$^+$ as part of a broader course on formal methods.[7] In addition, there are a few consultants who offer professional TLA$^+$ workshops.

According to some lecturers, some students may find the TLA$^+$ syntax challenging. Others have reported that grading assignments can be time-consuming.

5   Conclusion and Future Work

We have developed and regularly teach a two-day class on learning TLA$^+$ specifically for software engineers in the industry. The learning materials and verification tools for this class are freely available, and we have successfully used a cloud-based IDE that requires no installation and offers a sufficient free tier in many classes. According to our evaluations, our class has been well received and has helped expose more engineers to modeling and formal verification. However, increasing the adoption rate of these techniques remains a challenge.

In the future, we plan to adjust our surveys to more explicitly track student learning progress in order to better fine-tune and assess the impact of the class. Additionally, we hope to move away from using GitHub Codespaces, which requires a user account, and instead use a fully browser-based IDE. The *tla-web* project is a promising alternative that may also address the issue of the model checker's error messages being unclear or unhelpful.[8]

To address the challenges of teaching advanced concepts of temporal logic, particularly fairness, we are considering offering a 201 class on TLA$^+$ that uses *Raft* or *Paxos* as motivation. This class could also cover the symbolic model checker Apalache and the TLA$^+$ proof systems. In order to increase the adoption rate of TLA$^+$, we will experiment with extending the class to three days.

Acknowledgment. The author is grateful to Stephan Merz and Igor Konnov for their contributions to the teaching material and to the anonymous reviewers for their helpful feedback.

---

[7] https://github.com/tlaplus/awesome-tlaplus.
[8] https://github.com/will62794/tla-web.

Specs: AsyncTerminationDetection & EWD998

```
 1  ─────────────────── MODULE AsyncTerminationDetection ───────────────────
 2  EXTENDS Naturals

 3  CONSTANT N
 4  ASSUME NIsPosNat ≜ N ∈ Nat \ {0}

 5  Node ≜ 0 .. N − 1

 6  VARIABLES active, pending, terminationDetected
 7  vars ≜ ⟨active, pending, terminationDetected⟩

 8  terminated ≜ ∀ n ∈ Node : ¬active[n] ∧ pending[n] = 0

 9  TypeOK ≜ ∧ active ∈ [Node → BOOLEAN ]
10          ∧    pending ∈ [Node → Nat] ∧ terminationDetected ∈ BOOLEAN

11  Init ≜ ∧ active ∈ [Node → BOOLEAN ] ∧ pending ∈ [Node → Nat]
12         ∧    terminationDetected ∈ {FALSE, terminated}

13  Terminate(i) ≜ ∧ active' = [active EXCEPT ![i] = FALSE]
14              ∧ pending' = pending
15              ∧ terminationDetected' ∈ {terminationDetected', terminated'}

16  SendMsg(i, j) ≜ ∧ active[i]
17              ∧ pending' = [pending EXCEPT ![j] = @ + 1]
18              ∧ UNCHANGED ⟨active, terminationDetected⟩

19  Wakeup(i) ≜ ∧ pending[i] > 0
20           ∧ active' = [active EXCEPT ![i] = TRUE] ∧ UNCHANGED terminationDetected
21           ∧ pending' = [pending EXCEPT ![i] = @ − 1]

22  DetectTermination ≜ ∧ terminated ∧ UNCHANGED ⟨active, pending⟩
23              ∧ ¬terminationDetected ∧ terminationDetected' = TRUE

24  Next ≜ DetectTermination ∨ ∃ i, j ∈ Node :  Terminate(i) ∨ Wakeup(i) ∨ SendMsg(i, j)

25  Spec ≜ Init ∧ □[Next]_vars ∧ WF_vars(DetectTermination)

26  Stable ≜ □(terminationDetected ⟹ □terminated)
27  Live ≜ terminated ↝ terminationDetected

28  StateConstraint ≜ ∀ n ∈ Node : pending[n] < 3
29  ─────────────────────────────────────────────────────────────────────
30  ───────────────────────── MODULE EWD998 ─────────────────────────
31  EXTENDS Integers

32  CONSTANT N
33  ASSUME NIsPosNat ≜ N ∈ Nat \ {0}

34  Node ≜ 0 .. N − 1
35  Color ≜ {"white", "black"}

36  VARIABLES active, color, counter, pending, token
37  vars ≜ ⟨active, pending, color, counter, token⟩

38  TypeOK ≜ ∧ active ∈ [Node → BOOLEAN ] ∧ pending ∈ [Node → Nat]
39          ∧    color ∈ [Node → Color] ∧ counter ∈ [Node → Int]
40          ∧    token ∈ [pos : Node, q : Int, color : Color]
```

41 $Init \triangleq \wedge active \in [Node \to \text{BOOLEAN}] \wedge pending = [i \in Node \mapsto 0]$
42 $\wedge color \in [Node \to Color] \wedge counter = [i \in Node \mapsto 0]$
43 $\wedge pending = [i \in Node \mapsto 0] \wedge token = [pos \mapsto 0,\ q \mapsto 0,\ color \mapsto \text{"black"}]$

44 $InitiateProbe \triangleq \wedge token.pos = 0 \wedge \text{UNCHANGED } \langle active, counter, pending \rangle$
45 $\wedge (token.color = \text{"black"} \vee color[0] = \text{"black"} \vee counter[0] + token.q > 0)$
46 $\wedge token' = [pos \mapsto N - 1,\ q \mapsto 0,\ color \mapsto \text{"white"}]$
47 $\wedge color' = [color \text{ EXCEPT } ![0] = \text{"white"}]$

48 $PassToken(i) \triangleq \wedge \neg active[i] \wedge token.pos = i \wedge \text{UNCHANGED } \langle active, pending, counter \rangle$
49 $\wedge token' = [token \text{ EXCEPT } !.pos = @ - 1,\ !.q\quad = @ + counter[i],$
50 $!.color = \text{IF } color[i] = \text{"black" THEN "black" ELSE }@]$
51 $\wedge color' = [color \text{ EXCEPT } ![i] = \text{"white"}]$

52 $System \triangleq InitiateProbe \vee \exists i \in Node \setminus \{0\} : PassToken(i)$

53 $SendMsg(i) \triangleq \wedge active[i] \wedge \text{UNCHANGED } \langle active, color, token \rangle$
54 $\wedge\quad counter' = [counter \text{ EXCEPT } ![i] = @ + 1]$
55 $\wedge\quad pending' = [pending \text{ EXCEPT } ![\text{CHOOSE } r \in Node : \text{TRUE}] = @ + 1]$
56 $\wedge\quad \exists recv \in (Node \setminus \{i\}) : pending' = [pending \text{ EXCEPT } ![recv] = @ + 1]$

57 $RecvMsg(i) \triangleq \wedge pending[i] > 0 \wedge \text{UNCHANGED } \langle token \rangle$
58 $\wedge\quad active' = [active \text{ EXCEPT } ![i] = \text{TRUE}]$
59 $\wedge\quad pending' = [pending \text{ EXCEPT } ![i] = @ - 1]$
60 $\wedge\quad counter' = [counter \text{ EXCEPT } ![i] = @ - 1]$
61 $\wedge\quad color' = [color \text{ EXCEPT } ![i] = \text{"black"}]$

62 $Deactivate(i) \triangleq \wedge active' = [active \text{ EXCEPT } ![n] = \text{FALSE}]$
63 $\wedge \text{UNCHANGED } \langle pending, color, counter, token \rangle$

64 $Environment \triangleq \exists n \in Node : SendMsg(n) \vee RecvMsg(n) \vee Deactivate(n)$

65 $Spec \triangleq Init \wedge \Box[System \vee Environment]_{vars} \wedge \text{WF}_{vars}(System)$

66 $terminationDetected \triangleq$
67 $\neg active[0] \wedge token.pos = 0 \wedge token.color = \text{"white"} \wedge token.q + counter[0] = 0 \wedge color[0] = \text{"white"}$

68 $ATD \triangleq \text{INSTANCE } AsyncTerminationDetection$

69 $\text{RECURSIVE } Sum(\_, \_, \_)$
70 $Sum(f, from, to) \triangleq \text{IF } from = to \text{ THEN } f[to] \text{ ELSE } f[from] + Sum(f, from + 1, to)$

71 $Inv \triangleq$
72 $\wedge P0:: Sum(pending, 0, N - 1) = Sum(counter, 0, N - 1)$
73 $\wedge\quad \vee P1:: \wedge \forall i \in (token.pos + 1) .. N - 1 : \neg active[i]$
74 $\wedge \text{IF } token.pos\quad = N - 1 \text{ THEN } token.q = 0$
75 $\text{ELSE }\ token.q = Sum(counter, (token.pos + 1), N - 1)$
76 $\vee P2:: Sum(counter, 0, token.pos) + token.q > 0$
77 $\vee P3:: \exists i \in 0 .. token.pos : color[i] = \text{"black"}$
78 $\vee P4:: token.color = \text{"black"}$
79

References

1. Cousineau, D., Doligez, D., Lamport, L., Merz, S., Ricketts, D., Vanzetto, H.: TLA$^+$ proofs. In: Giannakopoulou, D., Méry, D. (eds.) FM 2012. LNCS, vol. 7436, pp. 147–154. Springer, Heidelberg (2012). https://doi.org/10.1007/978-3-642-32759-9_14
2. Dijkstra, E.W.: Shmuel Safra's version of termination detection (1987)
3. Konnov, I., Kukovec, J., Tran, T.H.: APALACHE: Abstraction-based parameterized TLA+ checker (2018)
4. Konnov, I., Kuppe, M., Merz, S.: Specification and verification with the TLA$^+$ Trifecta: TLC, Apalache, and TLAPS. In: Margaria, T., Steffen, B. (eds.) 11th International Symposium Leveraging Applications of Formal Methods, Verification and Validation (ISoLA 2022). Lecture Notes in Computer Science, vol. 13701, pp. 88–105. Springer, Rhodes (2022)

5. Kuppe, M.A.: The TLA+ Debugger. In: F-IDE 2022 (2022)

6. Kuppe, M.A., Lamport, L., Ricketts, D.: The TLA+ toolbox. Electr. Proc. Theor. Comput. Sci. **310**, 50–62 (2019). https://doi.org/10.4204/EPTCS.310.6

7. Lamport, L.: The temporal logic of actions. ACM Trans. Program. Lang. Syst. **16**(3), 872–923 (1994). https://doi.org/10.1145/177492.177726

8. Lamport, L.: Specifying Systems: The TLA+ Language and Tools for Hardware and Software Engineers. Addison-Wesley, Boston (2003)

9. Lamport, L.: The TLA+ Video Course (2017)

10. Yu, Y., Manolios, P., Lamport, L.: Model checking TLA$^+$ specifications. In: Pierre, L., Kropf, T. (eds.) CHARME 1999. LNCS, vol. 1703, pp. 54–66. Springer, Heidelberg (1999). https://doi.org/10.1007/3-540-48153-2_6

Teaching and Training in Formalisation with B

Thierry Lecomte[✉]

CLEARSY, 320 Avenue Archimède, Aix en Provence, France
`thierry.lecomte@clearsy.com`

Abstract. Despite significant advancements in the design of formal integrated development environments, applying formal methods in software industry is still perceived as a difficult task. To ease the task, providing tools that help during the development cycle is essential but proper education of computer scientists and software engineers is also an important challenge to take up. This paper summarises our experience of 20 years spent in the education of engineers, either colleagues or customers, and students, together with the parallel design and improvement of supporting modelling tools.

Keywords: B method · Training · Teaching

1 Introduction

Formal methods were developed to address software crisis by providing mathematically based techniques, including formal specification, refinement, proof, and verification. In theory, we now know how to use formal notations to write specifications and refine them gradually into a correct implementation, and use logic to prove programs correct. However, none of these techniques is easy to use by ordinary practitioners to deal with real software projects. The problem is the complexity of implementing formal methods and the scarcity of skilled labour. However, this difficulty can be alleviated by providing more suitable teaching content [6,7,10] and tools that facilitate the implementation of formal methods.

This paper presents the experience collected during the last 20 years, training (future) engineers to use the B method, while developing the IDE supporting the B Method (Atelier B), using them (the tool and the method) for industry strength projects (development, Verification & validation), and boosting their dissemination in academia by providing specific hands-on teaching sessions.

The article is structured in 7 parts. Section 2 introduces the terminology. Section 3 briefly introduces the main principles of the B method. Section 4 presents how our training and teaching activities are structured. Section 5 describes our teaching material. Section 6 presents the return of experience that we have collected over the last 25 years, before concluding.

© The Author(s), under exclusive license to Springer Nature Switzerland AG 2023
C. Dubois and P. San Pietro (Eds.): FMTea 2023, LNCS 13962, pp. 82–95, 2023.
https://doi.org/10.1007/978-3-031-27534-0_6

2 Terminology

This chapter clarifies a number of unusual terms and concepts used in this paper.

Atelier B is an industrial tool that allows for the operational use of the B Method to develop defect-free proven software[1].

B0 is a subset of the B language [2] that must be used at implementation level. It contains deterministic substitutions and concrete types. B0 definition depends on the target hardware associated to a code generator [4]. Most railways product lines use their own own specific code generator.

CSSP abbreviates CLEARSY Safety Platform. The CLEARSY Safety Platform is made up of a hardware execution platform, an IDE enabling the generation of diverse binaries from a single B model, and a certification kit describing its safety features as well as the safety constraints exported to the hosting system.

Safety refers to the control of recognized hazards in order to achieve an acceptable level of risk.

3 Introduction to the B Method

B [1] is a method for specifying, designing, and coding software systems. It covers central aspects of the software life cycle (Fig. 1): the writing of the technical specification, the design by successive refinement steps and model decomposition (layered architecture), and the source code generation.

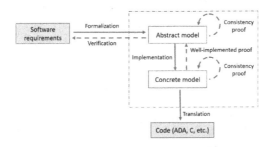

Fig. 1. A typical B development cycle, from requirements to code.

B is also a modelling language that is used for both specification, refinement (Fig. 2), and implementation (Fig. 3). It relies on substitution calculus, first order logic and set theory. All modelling activities are covered by mathematical proofs that finally ensure that the software system is correct.

B is structured with modules and refinements. A module is used to break down a large software into smaller parts. A module has a specification (called a machine) where a static and a dynamic description of the requirements are formalized. It defines a mathematical model of the subsystem concerned with:

---

[1] https://www.atelierb.eu/en/atelier-b-tools/.

- an abstract description of its state space and possible initial states,
- an abstract description of operations to query or modify the state.

This model establishes the external interface for that module: every implementation will conform to this specification. Conformance is assured by proof during the formal development process. A module specification is refined. It is re-expressed with more information: adding some requirements, refining abstract notions with more concrete notions, getting to implementable code level. Data refinement consists in introducing new variables to represent the state variables for the refined component, with their linking invariant. Algorithmic refinement consists in transforming the operations for the refined component. A refinement may also be refined. The final refinement of a refinement column is called the implementation, it contains only B0-compliant models. In a component (machine, refinement, or implementation), sets, constants, and variables define the state space while the invariants define the static properties for its state variables. The initialisation phase (for the state variables) and the operations (for querying or modifying the state) define the way variables are modified. From these, proof obligations are generated such as: the static properties are consistent, they are established by

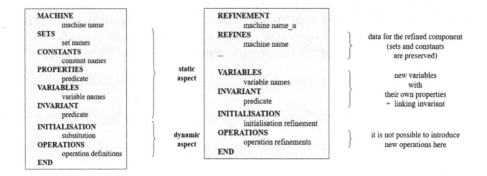

Fig. 2. Structure of MACHINE and REFINEMENT components.

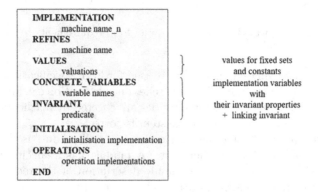

Fig. 3. Structure of IMPLEMENTATION component.

the initialisation, and they are preserved by all the operations. Atelier B contains a model editor merging model and proof by displaying the number of proof obligations associated to any line of a B model, its current proof status (fully proved or not) and the body of the related proof obligations.

Finally a B project is a set of linked B modules. Each module is formed of components: an abstract machine (its specification), possibly some refinements and an implementation. The principal dependency links between modules are IMPORTS links (forming a modular decomposition tree) and SEES links (read only transversal visibility). Sub-projects may be grouped into libraries. A software developed in B may integrate or may be integrated with traditionally developed code.

4 Training vs Teaching

Training and teaching are both aimed at delivering some pedagogical content to an audience. However objectives and expectations may vary between these two activities. This chapter presents the structure of these activities.

4.1 Training

As the software company responsible for the development of Atelier B, professional training has always been a key activity, be it to train colleagues or engineers from other companies. The objectives of the participants vary:

- [OBJ1] it may be to understand and analyse an existing B model when accepting a deliverable provided by a third party. This is a strong regulatory requirement when the deliverable contributes to a critical system. The need is to be able to read the models, to determine which properties are expressed and how they are distributed within the model.
- [OBJ2] The need is then to adapt a model without damaging the architecture. It is necessary to be able to write the required specifications and implementations in a correct and efficient way without calling into question the existing technical modelling choices. It is also necessary to preserve as much as possible the mathematical proofs of the model.
- [OBJ3] It may be a case of building *ex nihilo* a complete model B which corresponds to a given technical problem and which interfaces with particular software components. It is then necessary to know how to specify efficiently, how to distribute the processing within components and how to optimise the proof work through levels of refinement.
- As the B-method is definitely proof oriented, it is obvious that a model has to be developed in order to facilitate its proof. A model can be expressed in many ways and some of them are more easily proved by a theorem prover. [OBJ4] It is then necessary to have a deeper knowledge of automatic and interactive proof tools, of their capacity to prove such or such mathematical predicate.

Hence three training levels have been elaborated - "Understand B", "Practice B", and "Prove B" - to be practiced in this order and with some delay between each training to let the modeller assimilate the new concepts and get used to the technical environment.

"Understand B", directly aimed at [OBJ1], is designed to help understand the fundamental principles of the B Method and discover the B language. B is introduced as a method of formal specification and design with proof, which can go as far as the concrete level (with B0 language), and which offers formal specification and construction of a model by systematic description of its properties. Notions of modularity and hierarchy are presented, as a B model is built in a modular way, and its properties are introduced in a hierarchical way. Finally the proof is briefly presented as a mean to ensure the respect of invariant properties as it ensures in an exhaustive way that the code is in conformity with its specifications. To complete the picture, a description of the main uses of B in the industry is given. In a second part, predicates, mathematical expressions, and substitutions are all studied through their syntax and semantics, and implemented in short examples (often one-liners). The three types of B components (abstract machines, refinements and implementations) are presented. More than half of the training is hands-on session using Atelier B as a platform for experimenting the modelling and the automatic proof in B. The session is made of 4 consecutive full days, with a maximum number of 10 trainees for 2 trainers. Requirements for attending this training are a knowledge of the general principles of the development cycle of a system or software, a basic knowledge of computer science, and a mathematical knowledge at the level of a scientific baccalaureate.

"Practice B" covers both [OBJ2] and [OBJ3]. It is designed to help understand the principles of developing a B project, to practice building "good" B models, and to understand B language advanced concepts. Developing a B project requires to make clear the path leading from informal specifications to formal specifications. It also requires to know the modular construction of a B project and the various types of links between B modules, as well as the rules governing these links. A first methodological base is proposed on which to build a B project as an assembly of modules. The B model building practice is heavily based on exercises where the formal significance of "complying with specifications" is explained and linked to the proof obligations obtained. The participants are asked to create formal specifications on complete examples based on informal requirements. The principles for drafting models facilitating proof are studied. The session is similar to the "Understand B" one. At least, one month of intensive practice since "Understand B" training is required to let the participants increase their modelling skills.

"Prove B", directly aimed at [OBJ4], is designed to help verify models with proof, to understand how the automatic and interactive provers work. The verification activity relies on the use of an automatic prover to demonstrate most of the obligations of correct proof, the examination of remaining proof obligations to detect errors and the finalisation of the proof with the interactive prover. The automatic prover is described as a collection of collaborating proof strategies and mechanisms[2]. The main principles of the interactive prover are presented together with its interactive proof commands. Several methodological recommendations for a proper interactive demonstration allow to improve modelling efficiency. The session is similar to the ones above. At least, several months of intensive practice since "Practice B" training is required to let the participants increase their proof skills.

4.2   Teaching

Teaching at universities or engineering schools has a more pedagogical purpose than in a company. It is about educating students and complementing their ability to learn how to learn.

It uses a single resource base, which is made of:

- a presentation of the field of critical systems, which strongly recommends the use of formal methods for the highest criticality levels. The regulatory standards are introduced at this level.
- a presentation of the technical applications, the functions realised with the formal methods and the safety and security levels achieved.
- a modular presentation of the development cycle, the language and the associated tools.
- a corpus of simple examples for learning the language and more complicated (but simplified) examples from real systems.

The aim is to give students a formal touch, to teach them to model simple properties and to get to grips with the proof tool. In some cases, the code generation aspect is addressed. The teaching material is heavily based on the training resources. However the requirements are much lower than in a company and do not require a technical level to develop an industry-strength product.

5   Education Material

Most of education has been completed with traditional means such as slides for the 3 training levels[3] and books (pdf format)[4]. Teaching slides are directly offered to the students before the lectures, but are not released publicly. Exercises are completed on the blackboard or through computer manipulations.

Training "Understand B" comes with several exercises:

---

[2] The collaboration is static and has been designed decades ago to optimize proof benchmarks.

[3] https://b-method.gitbook.io/training-resources-for-atelier-b/b-training-course/slides.

[4] https://github.com/CLEARSY/CSSP-Programming-Handbook.

- **Specifying a resource management system** (model and proof obligations). 5 services have to be formally specified from a natural language description. For example, let the fourth service be named ReleaseResource. This operation takes a resource identifier rr as input, and may only be called when rr belongs to RESOURCES and indeed to the subset in_use as well. Its effect is such as available becomes available with element rr included, and in_use becomes in_use with element rr excluded. The resulting modelling is as follows:

```
ReleaseResource(rr) = PRE
    rr: RESOURCES &
    rr: in_use
  THEN
    available:=available\/{rr} ||
    in_use:=in_use-{rr}
  END
```

- **Simplified greatest common divisor**. The exercise makes use of the integer division to calculate the GCD of 2 positive integers that differ by no more than 2.
- **Batteries switch program**. 3 switches controlling 3 batteries powering a device have to be regularly controlled to avoid the same battery to discharge during a too long period. Properties are defined by learners (no short-circuit, power supply continuity) before their modelling.
- **Detection of the presence of two numbers in a list**. The exercise is aimed at using bool(P) expression.
- **Proving formal properties**: quantified predicates, function structure, simple induction. Several kinds of proof are introduced: contradiction, generalisation, and induction.
- **Block: Building a Complete Software B Project**. This software controls a railroad line, divided into fixed blocks. The purpose of the functionality is to establish safely, from the software point of view, which blocks are occupied by a train and which are free. Five different detectors are used but they are not accurate enough at the borders and they may be faulty. The project is made of 7 pre-existing components that need to be completed. For example, the operation set_block_occupancy should establish that a block having one of its border detector occupied or having its trackside detector occupied has to be occupied. In B, <: is the ASCII representation for set inclusion. A <: B means that the set A is included in the set B.

```
set_block_occupancy =
    BEGIN
        ob, tdl_alarm
            :(
            ob         <: t_block &
            tdl_alarm <: t_block &
            d_b2b[obd] \/ otd <: ob
```

```
    )
END
```

The exercise covers formal modelling of non-trivial properties[5], specification and implementation of operations, and provides a first experience of a multi-component B project. Exercises are often selected as they address concrete, well-known devices with a short specification, vague enough to generate discussion and to obtain various models.

Training "Practice B" comes with several other exercises:

- **Refinement principles.** Refinement proof obligations are studied, in particular the ones related to a missing gluing invariant.
- **Traffic light control system.** This is the occasion to find properties for this well-known system, from different points of view (safety, traffic-flow, end-user, maintenance). Several subjects are treated: linkage with an external environment, modular decomposition and maintainability.
- **Implementation concerns.** A collection of small examples related to ensuring the absence of overflow, an explosion of proof obligations, the proof of correctness for a simple loop, the SEES clause and avoidance of aliasing. There is also an introduction to abstract iterators[6] for loop.

```
cond, bl <-- iterate_t_block =
PRE
  blocks_to_treat /= {}
THEN
  ANY chosen_block WHERE
    chosen_block : t_block_i &
    chosen_block : blocks_to_treat
  THEN
    blocks_to_treat := blocks_to_treat - { chosen_block } ||
    treated_blocks := treated_blocks \/ { chosen_block } ||
    bl := chosen_block ||
    cond := bool(blocks_to_treat /= { chosen_block })
  END
END
END
```

- **Formal proof.** Several exercises to discover the proof activity: a proof of associativity (demonstration on paper then with the prover) and the language of proof-rules (introduction to the training "Prove B").

---

[5] Properties are not limited to typing. They require to use in combination diverse expressions and operators like composition, relational image, reverse, intersection, restriction in the domain, etc.

[6] With abstract iterators, the loop is prepared from the specification level by separating the iteration elements from the main substitution in a systematic way that could be efficiently implemented with automatic refinement. In the example below, t_block_i is the block super type.

- **Modelling access to an island through a tunnel.** Introduction to the Event-B modelling.

Training "Prove B" comes with a large collection of exercises, too large to be listed individually:

- **Modifying the model.** Adding ASSERTIONS to a model to trigger simplifications or proof mechanisms.
- **Understanding proof commands.** Introduction to most common interactive commands including Proof by cases, Set Solver.
- **Adding user rules.** Extend the mathematical rules database with user rules (that need to be validated by the tool or manually).
- **Ambiguity.** Some operators like - or * have several meaning types (set, integer). This ambiguity may block some simplification mechanisms. Adding hypotheses could solve the problem (command ah - *Add Hypothesis*). In the model below, assertions have to be demonstrated with invariant and properties as hypotheses. Assertions are ordered: assertion in line 232 comes as an hypothesis to assertion in line 233; assertions in lines 232 and 233 come as hypotheses to assertion in line 234.

```
CONSTANTS
  ii,jj
PROPERTIES
  ii: NAT &
  jj: NAT
ASSERTIONS
  ii-ii = 0;      /* ah(ii-ii = -ii+ii) */
  ii+1-1 = ii;    /* ah(ii+1-1 = -1+1+ii) */
  ii*jj = jj*ii   /* ar(CommutativityXY) */
END
```

These resources help understand the behaviour of the proof tool. Often the tool leaves you in the middle of a proof tree and it is up to you to figure out what is missing to continue/complete the proof. Directions are given to browse/discover the rules database, to write mathematical rules and proof tactics.

To complement these online resources, new formats have been made available:

- **videos**[7]: several videos demonstrating how to use Atelier B and the CLEARSY Safety Platform (CSSP).
- **MOOC**[8]: 20 videos covering the basic aspects of B. The examples come from [9]. 5 videos are related to B project management.
- **self-training document** (for colleagues only): a compilation of "Understand B" and "Practice B" with a small number of exercises.
- **Collections of models**[9]: a large number of models which allow the study of different styles of modelling in B.

---

[7] https://www.youtube.com/@atelierbclearsy.

[8] https://mooc.imd.ufrn.br/course/the-b-method.

[9] https://github.com/hhu-stups/specifications/tree/master/prob-examples/B.

6 Return of Experience

This chapter summarises our activity and our findings accumulated since the beginning of the training and teaching activity.

6.1 Activity

Training has been ensured during more than 25 years, mainly in Europe, for an audience ranging from junior to senior engineers, project managers, safety and security evaluators. Target industries include railways, smart card, automotive, nuclear energy, and telecommunications. All objectives (from [**OBJ1**] to [**OBJ4**]) have been addressed. Some participants followed the whole course, most of them were involved only in the first two training levels. Some sessions were specifically tailored for a particular kind of model or on an existing (difficult to complete or to maintain) B project. Indeed, some models are part of a critical infrastructure and have a life span of several decades. It is therefore necessary to maintain a level of competence that allows the associated software to survive company turnover.

Teaching has been ensured at various occasions during the last 25 years, on most continents: lectures in a university course, contribution to a doctoral school, tutorial or dedicated workshop for scientific conference, invited presentation. The duration varies from a few hours to 3 or 4 days, spread over a month. The audience is quite often composed of students in the last year of their master. The profiles varied greatly: future general engineers receiving an introduction to formal methods, students with training in mathematics, computer science, embedded systems or mechatronics, researchers, and teachers. The teaching has happened either as a standalone lecture or to complement a (more theoretical) lecture by a professor from the university or engineering school. In the latter case, the course was often asked to emphasise the industrial use of formal methods, with the course acting as a justification for academic teaching.

6.2 Feedback

Trainees vs Students. There is sometimes a huge difference between trainees and students. In industry, training is either carried out to address technical difficulties anticipated for the successful completion of a project, or is seen as a reward for professional performance. In almost all cases, the trainee is attentive and diligent during the training. This is not always the case for the student, for whom participation in the course may be compulsory because it is linked to a given curriculum whose content cannot be adapted. It therefore happens that the behaviour of these two populations (engineers, students) diverges significantly and that the students do not see the point of the course, even if industrial use cases are used as course material.

Handling Abstraction. Piaget [8] claimed that only one third of the population is able to handle abstraction. This proportion is somewhat reflected in our courses and training with:

- one or two people dealing with the questions faster than expected and getting ahead of the group in the practical work. These people, when recruited, make excellent practitioners;
- a first group understanding what is being done and why it is being done;
- a second group following the instructions given;
- a third group copying the results obtained by their peers or doing something else.

It should be noted that being a software developer does not imply a facility with formal modelling. Most developers do not have this ability, which is part of the reason why formal methods have difficulty being adopted in the industry when staff are selected solely for their availability and software skills. Our engineers are tested when they are recruited to see which group they belong to so that we don't make the mistake of assigning them to tasks that they will find very difficult to complete successfully.

Formal Models in Real Life. Demonstrating the value of formal methods for software development is difficult. You need to be able to learn and use a mathematical language effectively. It requires a willingness to make life difficult by adding properties to the software before it is built. Agile methods and the prior development of software demonstrators undermine this approach. Often the examples presented in the courses are simple (or even simplistic) and do not necessarily allow to apprehend the added value of formal methods. The industrial examples are too large and confidential to be able to provide this insight and convince the learners definitively. With the introduction of the CLEARSY Safety Platform for education, it is possible to bring formal methods closer to the real world. This programmable board allows to specify, implement, prove and execute control logic expressed with B that will interact with the outside world through a simple interface (digital inputs and outputs). For students in formal methods, it shows the concrete applicability of formal techniques to the real world. For students in computer science and embedded systems, it allows them to verify without testing a software development in exchange for an intellectual effort.

Teaching Proof is Difficult. Proof has always been the stumbling block for teaching B. It is rather easy to explain how to model behaviour and properties. It is much more difficult to try to understand why certain points are not automatically proven[10]. This understanding has to be done through the prism of

---

[10] For cost reasons, the development of the core of the prover was frozen in 1998 to avoid that prover evolutions generate regressions of proof. Indeed, to interactively prove a proof obligation costs about 35 euros. The modification of the proof status of an industrial project following the evolution of the prover could impact tens of thousands of proof obligations and would not be acceptable to the industrialist.

multiple proof tools (theorem provers, solvers, model-checkers), which requires proven skills in mathematical proof (and an appetite for the subject). One must be able to determine whether the lack of automatic proof is due to:

- a limitation of the tool - then one must determine whether to modify the model to make it more provable by reformulating the properties and behaviour, or by using certain commands of the proof tool;
- a modelling error - then the model must be modified.

The proof activity is intimately linked to the modelling activity and to be effective must be carried out by the same person. The integration of the proof status in the model editor (Fig. 4) allows the modeller to be aware of the complexity of his modelling in terms of proof. This complexity could be quickly estimated based on the number of proof obligations, their localization and their automatic proof rate. The number of remaining proof obligations is a good measure of the complexity of a model, that has to be confirmed by visual inspection. The connection of Atelier B to external provers has and will improve the automatic proof rate. The difficulty of proof is thus reduced but not eliminated.

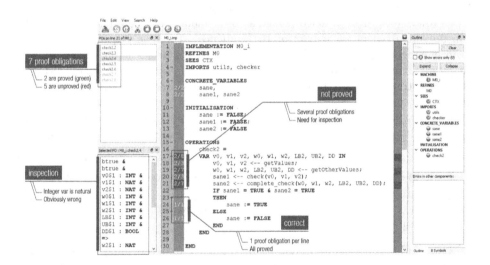

Fig. 4. Atelier B model editor showing proof status.

Automating Refinement. Refinement is at the heart of B. It is a hard point when it comes to transforming non-trivial abstract structures and substitutions into their implementation in one or more steps. The refinement techniques depend on the human modellers and their experience. For the development of the Meteor metro safety automation, MATRA Transport [5] has developed and documented refinement techniques to systematise their use. This resulted in an automatic refinement tool that was later redeveloped for Atelier B. This tool (BART) automates the refinement of a B machine, using an extensible base of refinement rules and an inference engine to apply these rules to an abstract B model. The tool refines the data and then the operations through a process:

- automatic: the engine applies its refinement rule base to the abstract model. It stops when no rule can be applied to the structure/substitution being refined.
- interactive: the modeller must therefore complete the refinement rule base and then restart the automatic process. Refinement is complete when the machine has been successfully transformed.

Neither the tool nor the rules are proven: if the refinement produced is incorrect, it will not be provable. This tool was introduced in the mid-2000s in the hope of making refinement more easily accessible. The expected effect has not been achieved because in fact this tool, which allows the automation of a refinement process, requires a great deal of expertise on the part of the operator, who must know how to refine and model his knowledge in the form of rules.

7 Conclusion and Perspectives

Either training or teaching B are activities difficult to complete satisfactorily. The subject (set theory, first order logic, refinement, proof, etc.) is difficult and can only really concern a part of the audience. Our teaching resources have been enhanced over 25 years to address primarily a professional audience and therefore focus on the modelling of concrete problems/systems. In the meantime, the supporting tools have been improved: the editor integrates proof information while third party provers extend the proof system. New video-based resources have been available while the animation (graphic or not) of models was promoted [3]. Several on-going research projects are aimed at easing the proof process, to make the B modelling more appealing:

- With the project AIDOART[11], Artificial Intelligence could ease the proof process by suggesting proof tactics.
- With the projects BLASST[12] and ICPSA[13], third party provers/solvers could improve proof performances.

One could also imagine having a Big Data based tool offering modelling choices like copilot/Github using OpenAI. In fact, any technological innovation that would simplify the application of the B-method would be welcome to promote learning.

Acknowledgements. The work and results described in this article were partly funded by ECSEL JU under the framework H2020. as part of the project AIDOaRt (AI-augmented automation supporting modelling, coding, testing, monitoring and continuous development in Cyber-Physical Systems).

---

[11] https://www.aidoart.eu/.
[12] Enhancing B Language Reasoners with SAT and SMT Techniques.
[13] Interoperable and Confident Set-based Proof Assistants.

References

1. Abrial, J.: The B-book - Assigning Programs to Meanings. Cambridge University Press, Cambridge (2005)
2. Abrial, J.: Modeling in Event-B - System and Software Engineering. Cambridge University Press, Cambridge (2010)
3. Bendisposto, J., et al.: PROB2-UI: a java-based user interface for ProB. In: Lluch Lafuente, A., Mavridou, A. (eds.) FMICS 2021. LNCS, vol. 12863, pp. 193–201. Springer, Cham (2021). https://doi.org/10.1007/978-3-030-85248-1_12
4. Boulanger, J.: Formal Methods: Industrial Use from Model to the Code. Wiley, New York (2013)
5. Burdy, L., Meynadier, J.M.: Automatic refinement. In: FM 1999 workshop - Applying B in an Industrial Context : Tools, Lessons and Techniques, Toulouse, France, Proceedings (1999)
6. Cataño, N.: Teaching formal methods: lessons learnt from using event-B. In: Dongol, B., Petre, L., Smith, G. (eds.) FMTea 2019. LNCS, vol. 11758, pp. 212–227. Springer, Cham (2019). https://doi.org/10.1007/978-3-030-32441-4_14
7. Istenes, Z.: Experiences of teaching formal methods in higher education. Formal Methods in Computer Science Education (FORMED 2008) (2008)
8. Kramer, J.: Is abstraction the key to computing? Commun. ACM **50**, 36–42 (2007)
9. Schneider, S.: The B-method (Cornerstones of Computing). Macmillan Education, UK (2001)
10. Zhumagambetov, R.: Teaching formal methods in academia: a systematic literature review. In: Cerone, A., Roggenbach, M. (eds.) FMFun 2019. CCIS, vol. 1301, pp. 218–226. Springer, Cham (2021). https://doi.org/10.1007/978-3-030-71374-4_12

Teaching Low-Code Formal Methods with Coloured Petri Nets

Somsak Vanit-Anunchai[(✉)]

School of Telecommunication Engineering, Institute of Engineering,
Suranaree University of Technology, Muang, Nakhon Ratchasima, Thailand
somsav@sut.ac.th

Abstract. This paper proposes teaching formal methods using Coloured Petri Nets and CPN Tools. This tool can hide the mathematical complexity and provides a simple and easy way for non-technical users to build their models without extensive coding (low-code). According to our experience in Coder Dojo workshops, when applying CPN Tools to a certain class of simple problems, even the youths can create, maintain the model and play with it. They can use the formal methods without realizing it. This paper tries to convince the idea of low-code formal methods by illustrating various examples. Finally, we suggest teaching low-code formal methods not as a separate subject but rather weaving it into the mainstream curriculum.

Keywords: Game24 · No code platform · Coder Dojo · PIPE · CPN tools

1 Introduction

Coder Dojo is a volunteer-led community of practice holding free programming workshops for the youths. During one of the Coder Dojo workshops in 2016, a ten years old boy challenged me to compete with him in the "Game24". The Game24 is a game giving four integer numbers. For each player, the goal is to apply three arithmetic operations; either additions or subtractions or multiplications or divisions; in order to get the result of 24. This game sometimes has no solution. To check whether the solutions were exist or not, I used a formal method tool, Coloured Petri Nets (CPN) and CPN Tool, to create the Game24 model. It took me only 20 min to build and test the model; another 1 min to generate the state space and query the answers. I was intrigued when learned that other programmers spent hours or days creating this game in Java or C. This ignited me with many questions.

1) Why is the Game24 model using CPN Tools a lot simpler than using other computer languages?
2) What kinds of other games or applications can we get similar benefits from Petri Net formalism?

C. Dubois and P. San Pietro (Eds.): FMTea 2023, LNCS 13962, pp. 96–104, 2023.
https://doi.org/10.1007/978-3-031-27534-0_7

3) Should we introduce Coloured Petri Nets or formal methods to the youths in primary school or secondary school? If yes, what should we teach them?

This paper attempts to address the answers to these three questions. The rest of this paper is organised as follows. Section 2 briefly explains the CPN model of the Game24. Section 3 discusses the idea of no-code/low-code formal methods. Section 4 suggests where the low-code formal methods should be used. Section 5 outlines a graduate course on the formal methods that I teach. Section 6 discusses the related work. Section 7 presents conclusions and suggests future work.

2   The Coloured Petri Net Model of the Game24

Coloured Petri Net (CPN) [5,6] is a formal method tool widely used for modelling and analysing complex concurrent and distributed systems. It uses graphical notation and abstract data structures giving conciseness together with a high level of expressiveness. We used CPN Tools [3] to create, edit, simulate and analyse the model.

Figure 1 illustrates the model of the Game24 and its result. At the initial marking, there are four integer tokens; 5, 6, 7, and 9; in Place *Start*. The CPN model takes two tokens assigned to variables; a and b; performs an arithmetic operation and puts the result back in Place *Start*. Then the number of tokens reduces to three. The model continues another two arithmetic operations until one token is left in Place *Start*. This execution is only one of many possible scenarios. After we generate every possible scenario using the state space tool, we search for the dead markings that have the number 24 in Place *Start* using Standard ML (SML) code shown in Fig. 1 b). In this particular example, we got only one answer, $(7-5) \times 9+6$. Place *RECORD* typed by *REC*; the product of two integers and one string; is used to keep track of the operation leading to the result. Without Place *RECORD* the model still works properly.

The CPN model of Game24 is simple because, firstly, it amounts to little code. Secondly, the main task of Game24 is to try arithmetic operations in every possible scenario and search for the answers. While the others have to write their search using Java or C, CPN Tools already provides us the state space generation and state space search. Thirdly, because it is a mathematical problem, the specification and analysis are already formal. Thus, building the formal model from informal specification and conducting formal analysis are straightforward, not ambiguous tasks.

3   No-Code/Low-Code Paradigm

Typically in a no-code platform users with no knowledge of how coding works can build their own applications by themselves while low-code development addresses a simple and easy way for non-technical users to build their applications without extensive coding. In some applications such as Fig. 1, we can use CPN Tools to create the Game24 model with very few lines of SML codes. The competence in

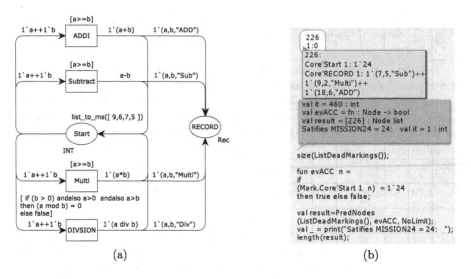

Fig. 1. (a) The CPN diagram of the Game24. (b) After generating the state space, searching the result of the game using query language (Standard Meta Language).

SML definitely helps enhance the usage of CPN Tools. However, CPN Tools has a simplified version of SML, called "CPN ML" so that non-technical users can use CPN Tools with ease. Thus, we consider CPN Tools as the low-code platform. Moreover, we introduce Place/Transition Nets (P/T Nets) to the youths in Grades 4–6 who use PIPE (Platform Independent Petri Net Editor) [8] as the no-code platform. Even though these no-code and low-code platforms have major drawbacks in diversity, customization, and scalability. Because of their simplicity, they are excellent choices being used for teaching formal methods to children aged 10 to 15.

Learning by Examples. Kids love games and riddles. They like playing and competing with each other in games. They also would like to create and modify games by themselves. Here are the first three games that Coder Dojo students have addressed using PIPE to create the no-code P/T Net model.

The Stone Picking Puzzle: Two players pick 1, 2, 3 or 4 stones from a stack of 20 stones alternatively. The player who picks the last stone loses. The students used the P/T Net models as a toy. They can also change the rules. For example, whoever picks the last stones win, or at each turn the player cannot pick two stones. The P/T Net model of the Stone Picking puzzle is illustrated in Fig. 2 a).

The Egg Fusion Puzzle: Player A has got the black eggs. Player B has got the blue eggs. And Player C has got the red eggs. A black egg fusing with a blue egg becomes a red egg. A red egg fusing with a blue egg becomes a black egg. And a

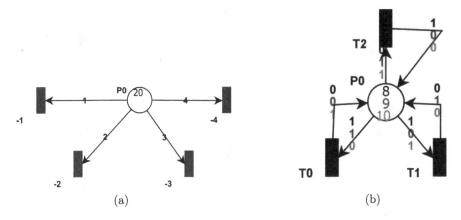

Fig. 2. (a) The P/T net model of the stone picking puzzle. (b) The P/T net model of the egg fusion game.

red egg fusing with a black egg becomes a blue egg. Each player puts a number of eggs into the oven. After all fusions, whoever has the eggs left in the oven is the winner. The P/T Net model of the Egg Fusion puzzle is illustrated in Fig. 2 b). This egg fusion model introduces the new concept of coloured tokens to the students.

The River Crossing Puzzle: A man brings with him a goat, apples, and a wolf. On the way, he must cross a river. The boat is very small and he can take only one of his belongings. Without his presence, the goat will eat the apples and the wolf will eat the goat. Only the man can row the boat. How can he get everything across the river? The no-code P/T Net model of this River Crossing Puzzle is illustrated in Fig. 3. An actor in this puzzle is modeled by the place with a token. Four inhibitor arcs are used to stop the goat from eating the apples and stop the wolf from eating the goat when the man is present. When there are more actors, the arcs in the model become messier. Then we introduce Coloured Petri Nets where many places can be folded into a single place in which actors are modeled by complex data structure tokens. Transitions can also be folded using variables on the arcs. Figure 4 shows the Coloured Petri Net version of the P/T model in Fig. 3. The state space of the River Crossing Puzzle comprises 21 states as shown in Fig. 5. The initial state is node 1. The solutions to the problem are the state sequences 1, 3, 7, 10, 13, 17, 19, 20 and 1, 3, 7, 11, 15, 17, 19, 20.

Discussion. From my observation during the Coder Dojo workshops, children with no background in coding can grasp very quickly the idea of P/T Nets and Coloured Petri Nets. They can understand the concept of concurrency and divided-and-conquer. They can create and modify the model very quickly and enjoy playing the simulation. The students are encouraged to find alternative solutions. However, we still had troubles with teaching reachability analysis. Although I emphasized that the reachability analysis easily reveals the solutions

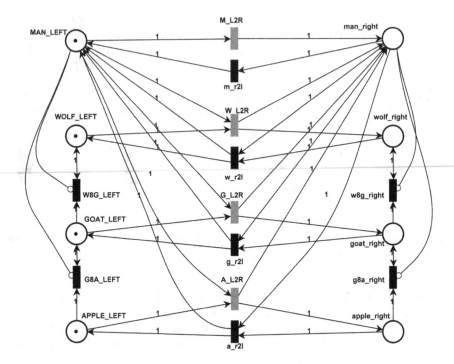

Fig. 3. The P/T net model of the crossing river puzzle.

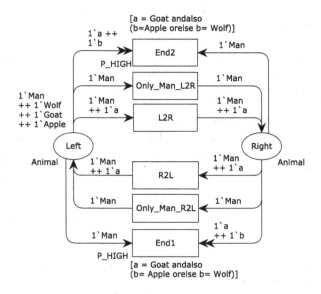

Fig. 4. The coloured petri net model of the crossing river puzzle.

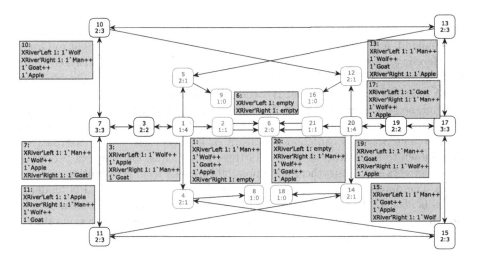

Fig. 5. The state space of the crossing river puzzle.

to the puzzle, only the high school students could comprehend it. Perhaps it is too difficult and requires a lot of concentration for young children.

Before the pandemic, our Coder Dojo students modeled many game puzzles. These puzzles have mixed features among moving objects, constraints, and logic. Nevertheless, mathematical problems such as searching the prime numbers or permutations/combinations are easily modeled using CPN Tools as well. Their solutions can also be computed using the state space tool easily.

4   Weaving Low-Code Formal Methods into the Mainstream Curriculum

Nowadays there is debate in Thailand about which year children should start learning computer programming. Many parents believe that computer programming does not require mathematics and have their children learning Scratch and Python since Grade 4. They also suggest that computer programming helps their children understand mathematics. The conservatives believe that computer programming does require mathematics. The children should be mature enough to learn computer programming when they are in Grade 10. However, according to my experience with Coder Dojo's students, young children can learn concepts of formal methods through games and problem-based learning using appropriate tools. Instead of teaching them traditional formal methods as separate subjects, we should weave low-code formal method tools into other subjects where possible. Here are two examples illustrating that CPN Tools can be useful. The first CPN model is an exercise of grade 10 combinatorial problems. The second example comprises six CPN models that are used to teach the "Principle of Reliable Data Transfer" in the computer networking course.

4.1   Combinatorial Problems

In Thailand, a branch of mathematics called "Combinatorics" is taught in Grade
10. Here is an example that a Coder Dojo student modeled and analysed using
CPN Tools. Six balls in a bag, three white, one yellow, one red, and one blue, are
given to students named A and B. Students either get some balls or no ball. In
the end, no ball is left in the bag. How many possibilities are there? Using Grade
10 maths, the yellow ball can choose either A or B. The red ball can choose either
A or B. The blue ball can choose either A or B. Then, there are 8 alternatives.
For the white ball, either A or B gets three white balls. Or one gets two while
the other gets one. Thus, there are 4 possibilities. The total possibilities are 8×4
$= 32$. The number of dead markings that are generated from the CPN model in
Fig. 6 is also 32. Looking into the state space offers a better understanding of
how to solve combinatorial problems and verify the answers at the same time.

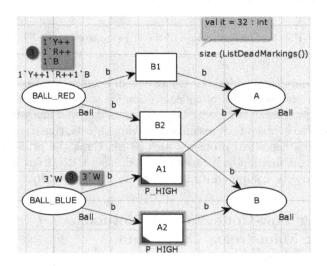

Fig. 6. The CPN model of combinatorial problem.

4.2   Principle of Reliable Data Transfer

In the Computer Networking textbook [7] Kurose and Ross use automata to
explain the *Principle of Reliable Data Transfer*. They start with an automaton
modelling an ideal channel, then introduce a channel imperfection. Using the second automaton, protocol mechanisms that solve the problem are then proposed
and explained. After that, the next defect and its solution are introduced. There
are 6 automata in this topic. Modifying from [6], we translated each automaton
into a CPN model. After spending half an hour to understand how Coloured
Petri Nets operates, all graduate students agreed that the CPN models provided clearer and better understanding than the automata did. Figure 7 shows
one of the six CPN models we used in teaching our graduate course, computer
networking.

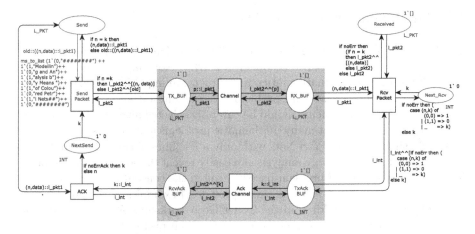

Fig. 7. The CPN model of the stop and wait protocol.

5   System Design Technique

I have taught a course titled; "System Design Technique" at Suranaree University of Technology (SUT) for graduate students since 2009. The objective of the class is to prepare telecommunication engineering students in order to conduct research on protocol verification based on the approach illustrated in [1]. Because telecommunication engineering students do not have background in discrete mathematics and automata, their curriculum is more practical and low-code formal methods rather than proof and extensive programming. The course is divided into three parts. The first part (3 weeks) starts with the introduction to Discrete-time Event Systems; Automata and Language [2]. These are important foundations for protocol verification. The second topic (2 weeks) is P/T Nets which the students learn by doing examples using PIPE. The graduate students seem to be interested in real-world applications rather than solving puzzles. The third part (4 weeks) is about Coloured Petri Nets and CPN Tools [6]. Another three weeks are spent on a term-project.

6   Related Work

The idea of weaving the use of formal methods into the existing curriculum is not new. Wing [9] listed the concepts and tools of the formal methods that we should teach and how to embed them into the existing curriculum. The idea of teaching the formal methods to the youths is also not new. Gibson [4] demonstrated that teaching young children (aged 7) formal methods concepts through games and problem-based learning was possible. He believed that this was a good way to introduce computer science in schools. He also suggested that mathematics is important for computer science education meanwhile computer science can help teach mathematics.

7   Conclusion and Future Work

Teaching formal methods to computer science students is hard but teaching them to non-computer science students is harder. Not only because they lack foundations in discrete mathematics and programming skill but also because the context of the exercises and problems are not interesting enough for them. Thus, examples and problems should be carefully selected according to students' interests. We discover that many mathematic problems and puzzles can be easily modeled using Coloured Petri Nets with only a few lines of code. The important advantage is that state space analysis can easily reveal the solutions to the problems. With low-code and problem-based learning, the concept of formal methods can be taught to the youths. This paper provides 4 examples of the puzzles that are used to teach the children in the Coder Dojo workshop. Another two examples, the combinatoric problem and the stop and wait protocol, demonstrate how the low-code formal methods can be used to help students study mathematics and computer networks.

In the future, there are two extensions that we would like to pursue. First, developing the animation to support CPN tools. Second, interfacing CPN models to real-world devices. These two extensions will help to teach not only formal methods but also the subject that the students study.

References

1. Billington, J., Gallasch, G.E., Han, B.: A coloured petri net approach to protocol verification. In: Desel, J., Reisig, W., Rozenberg, G. (eds.) ACPN 2003. LNCS, vol. 3098, pp. 210–290. Springer, Heidelberg (2004). https://doi.org/10.1007/978-3-540-27755-2_6
2. Cassandras, C.G., Lafortune, S.: Introduction to Discrete Event Systems. Springer, New York (1999). https://doi.org/10.1007/978-0-387-68612-7
3. CPN Tools home page. http://cpntools.org
4. Gibson, J.P.: Formal methods : never too young to start. In: FORMED 2008 : Formal Methods in Computer Science Education, pp. 149–158, Budapest, Hungary, March 2008
5. Jensen, K., Kristensen, L.M.: Colored petri nets: a graphical language for formal modeling and validation of concurrent systems. Commun. ACM **58**(6), 61–70 (2015)
6. Jensen, K., Kristensen, L.M.: Coloured Petri Nets: Modelling and Validation of Concurrent Systems. Springer, Heidelberg (2009). https://doi.org/10.1007/b95112
7. Kurose, J.F., Ross, K.W.: Computer Networking: A Top-Down Approach. Pearson, London (2013)
8. PIPE2 home page. https://pipe2.sourceforge.net/
9. Wing, J.M.: Invited talk: weaving formal methods into the undergraduate computer science curriculum (Extended Abstract). In: Rus, T. (ed.) AMAST 2000. LNCS, vol. 1816, pp. 2–7. Springer, Heidelberg (2000). https://doi.org/10.1007/3-540-45499-3_2

Author Index

C. Dubois and P. San Pietro (Eds.): FMTea 2023, LNCS 13962, p. 105, 2023.
https://doi.org/10.1007/978-3-031-27534-0

Printed in the United States
by Baker & Taylor Publisher Services